Lecture Notes
in Control and Information Sciences 307

Editors: M. Thoma · M. Morari

S.J. Kwon · W.K. Chung

# Perturbation Compensator based Robust Tracking Control and State Estimation of Mechanical Systems

With 75 Figures



Springer

## Authors

Dr. SangJoo Kwon
Division of Applied Robot Technology
Korea Institute of Industrial Technology
Ansan 425-791
Korea

Prof. Wan Kyun Chung
National Research Lab on Intelligent Mobile Robot
Pohang University of Science and Technology (POSTECH)
Pohang 790-784
Korea



ISSN 0170-8643

ISBN 3-540-22077-1 Springer Berlin Heidelberg New York

Library of Congress Control Number: 2004107832



Springer is a part of Springer Science+Business Media

springeronline.com



Typesetting: Data conversion by the authors.
Final processing by PTP-Berlin Protago-TeX-Production GmbH, Germany
Cover-Design: design & production GmbH, Heidelberg
Printed on acid-free paper 62/3020Yu - 5 4 3 2 1 0

To my family: Miyoung, Sunmok, and Sunlin
– SangJoo Kwon

*But God commendeth his love toward us, in that,
while we were yet sinners, Christ died for us. (Romans 5:8)*

# Preface

As a practical way to achieve robust motion control and state estimation (Kalman filtering), a promising approach in terms of the perturbation compensator (or perturbation observer equivalently) is investigated in this book, where the *perturbation* means lumped uncertainty not considered in the nominal plant model based on which the feedback controller and state estimator are designed. The perturbation compensator can be regarded as a kind of model regulator which drives the physical plant with uncertainty to the nominal model. Specifically, in mechanical systems where the nonlinear friction is regarded as a dominant disturbance, the perturbation observer actually operates as an adaptive friction compensator. Fundamentally, it is a decoupling control approach in contrast to the multivariable control method based on the linear optimal control solutions. Above all, the perturbation observers are physically intuitive, structurally simple, and easily implementable. Hence, these can be readily accepted by industry to increase the performance of many control systems. Based on a unified view on the existing class of perturbation observers, some novel approaches for design and analysis of perturbation observer are investigated in this book and they are extended to robust motion control and robust state estimation (Kalman filtering also). The stability and sensitivity characteristics of the perturbation observer loop is made clear through discrete-time analysis. Here, the main results are given chapter by chapter.

In Chapter 2, a robust tracking control method is proposed by combining the perturbation compensator which effectively attenuates plant perturbations including model uncertainty and external disturbance with the nominal tracking controller designed using the sliding surface. This approach enables a smooth sliding mode in tracking control loop without chattering problem. An unified view is given on a class of perturbation observers and three kinds of equivalent expressions for the perturbation of a plant is described. In terms of the equivalents, we propose the feedfoward perturbation observer (FFPO), the feedback perturbation observer (FBPO), and the sliding mode perturbation observer (SMPO). Successively, by hierarchically adopting these three observers to attenuate the *residual perturbation*, the hierarchical perturbation compensator (HPC) is constructed. The stability and robustness property of HPC is clarified through the analysis of error dynamics. Adaptive and integral property of HPC greatly enhances the tracking performance with minimal control effort. The issue of actuator saturation is also considered.

In Chapter 3, the concept of residual perturbation and hierarchical perturbation compensation in Chapter 2 is extended to general $n$-loop case. As a result, the multiloop perturbation compensator (MPEC) is formulated. A robust stability condition with respect to the inertia variation is derived as a function of the number of compensation loops. In MPEC, as the number of compensation loops increases, the external disturbance condition for system stability is gradually relaxed and the perturbation attenuation performance is remarkably enhanced but the robust stability margin on the modeling error becomes more strict. By combining MPEC with nominal feedback controller, a robust motion controller is synthesized and it is applied to XY positioner and robot manipulator.

In Chapter 4, the perturbation observer (compensator equivalently) is investigated in discrete-time viewpoint. In fact, the design of the low pass filter has been a central issue in the class of perturbation observers, for example, disturbance observer (DOB). However, the effect of the low pass filter (so-called $Q$-filter) on the performance and robustness of the perturbation observer has not been clarified yet, specifically in discrete-time domain. In this chapter, this problem is answered considering the general order of discrete $Q$-filter and the inertia perturbation as a structured model uncertainty. The results illustrate well how the performance and robustness of the perturbation observer involved compensation loop is changed in discrete-time domain according to the parameter variation of $Q$-filter and plant. Accordingly, some guidelines on the $Q$-filter selection for the trade-off between performance and robustness are suggested.

In Chapter 5, a combined state estimator–perturbation observer is synthesized by unifying standard linear state estimator (Luenberger observer) and perturbation observer. It enables robust state estimation for uncertain dynamical systems and simultaneously provides full-state to the perturbation observer under output feedback condition. The combined observer can be regarded as a practical state estimator to overcome the limit of the imperfectness of plant model. It has the merit that it is given as a recursive discrete-time form and above all it requires no knowledge of plant uncertainty. A coupled estimation error dynamics is derived and the related technical issues such as stability and noise sensitivity are addressed. The combined observer setting is also extended to stochastic systems and the discrete Kalman filter is reformulated by including the perturbation estimate update process. Numerical examples and experimental results validate the practicality of the combined observer and the modified Kalman filter.

In Chapter 6, some control problems related to the coarse/fine dual-stage are included. First, a dual-stage control architecture is suggested, where the perturbation compensator is necessitated to guarantee a good tracking performance in spite of plant uncertainty. Also, the minimum-time control using the coarse/fine dual-stage is discussed. The fast and fine positioning capability of the coarse/fine dual-stage is shown through the microteleoperation issue.

I am grateful to many people who made contributions to this book directly or indirectly. Special thanks to Professor Wan Kyun Chung who was my advisor, Director of National research lab on Intelligent mobile robot and Professor Youngil Youm, Director of Robotics and bio-mechatronics lab in Pohang University of Science and Technology (POSTECH), for supporting me financially and morally through my doctoral course. I would also like to thank other committee members for my Ph.D. dissertation: Professor Jin Soo Lee in POSTECH, Professor Ilhong Suh in Hanyang University, and Professor Masayoshi Tomizuka in University of California at Berkeley, for their sincerity and familiarity. Acknowledgement is given to "POSTECH Robot family" and colleagues in Korea Institute of Industrial Technology. Finally, this book is dedicated to my family, source of my happiness.

Mar. 2004
Ansan, Korea

SangJoo Kwon

# Contents

# 1

# Introduction

## 1.1 Motivation of the Book

The advanced optimal control methods such as LQG/LTR control [16,59,81], $H_\infty$ control [59,67], and $\mu$-synthesis [59] enable to systematically design stabilizing controllers which meet robust stability and performance on the plant uncertainty. In these control frameworks, however, the plant uncertainty is usually represented by the unstructured uncertainty such as multiplicative uncertainty or additive uncertainty which are lacking in phase information and whose upper bound of magnitude is assumed to cover the worst case of plant uncertainty. Thus, it inevitably includes a class of plants which may practically never happen. As a result, these controllers usually have high-gain characteristics and make a control system very conservative in performance respect although they guarantee robust stability for the assumed plant uncertainty. In fact, these schemes are scarcely in practical use in industry in spite of the massive research works through the past twenty years. The disregard from the industry seems to be caused by the facts that the robust optimal control methods do not give physical assurance in spite of the mathematical completeness and the industrial engineers prefer simple control algorithms considering implementation cost and hardware maintenance.

As a practical way to perform robust motion control, the perturbation compensator approach will be investigated in this book, where the term of *perturbation* means the lumped uncertainty of a plant excluded in nominal model based on which a feedback controller is designed. The perturbation compensator (or perturbation observer equivalently) can be regarded as a *model regulator* which drives the physical plant with uncertainty to the nominal model. Specifically, in mechanical systems where the nonlinear friction is a dominant disturbance, the perturbation observer actually operates as an adaptive friction compensator. Basically, it is a decoupling (or decentralized) control approach in contrast to the multivariable control based on the linear optimal control solutions. Above all, the perturbation observers are physically intuitive, structurally simple, and easily implementable. So, these can be readily accepted by industry to increase control system performance. In this book, based on a unified view on the existing class of perturbation observers, some novel approaches for the design and analysis of the perturbation observer are suggested and they are extended to *robust motion control* and *robust state estimation (or robust Kalman filtering)*. Specifically, the relationship between the motion control performance and the perturbation observer design parameters is made clear through the discrete-time design and analysis. Moreover, we stress experimental verifications of the proposed schemes.

On the other hand, the sliding mode control [5,85,107] could be a powerful way to achieve robust control performance for uncertain systems including linear or nonlinear systems. But, the control chattering which is an intrinsic phenomenon in variable structure controllers has been a hot issue. To remedy this problem, a number of smoothing techniques, for example [68, 85, 107], were suggested. However, the application to real systems would be inevitably restricted as far as the discrete switching input over the bandwidth of control actuator is kept in the sliding mode control. In Chapter 2, it is claimed that the discrete switching input of the conventional sliding mode control could be abandoned if we adopt the perturbation compensator in the sliding mode control. In the suggested *perturbation compensated sliding mode control* (PC-SMC), the perturbation compensator is substituted for the discrete switching input to address robustness to the plant uncertainty. However, the robustness due to the switching control can be well recovered by the perturbation compensator.

These days, fine actuators such as piezoelectric actuator with extremely small travel range but nano-level resolution are popularly used for motion control applications such as micro-teleoperation. Another interest of this book is taken in the coarse/fine dual-stage compound actuation system using both coarse and fine actuator, which is a convincing solution to overcome the performance limit of many positioning systems equipped with electrical motors or hydraulic actuators. In Chapter 6, some control strategies including null-motion control and robust time-optimal control are investigated to best utilize the capability of the coarse/fine dual-stage.

## 1.2 Issues in Mechanical Systems Control

In mechanical systems for industrial applications, the control performance is greatly affected by diverse uncertainties which cannot be easily identified. It includes such as nonlinear friction, inertial cross coupling, flexible mode, and also the hard nonlinearities such as backlash, hysteresis, and actuator saturation. Furthermore, in many cases, the control system is exposed to unpredictable external disturbances. For example, XY positioning table in the CNC machining center are persistently influenced by the cutting force and the robot manipulator in pick and place tasks may undergo abrupt inertia variation. Hence, the productivity of industrial manufacturing systems will be largely dependent upon how intelligently the servo controller is able to compensate these adverse effects. The presence of uncertainties in the control system makes the fine positioning or tracking very difficult but also challenging.

The sensitivity of the plant behavior with respect to the uncertainties can be considerably reduced by applying a simple feedback like the PID control and the control performance could be quite satisfactory in many cases. However, it must be limited to the tasks in which the simple control is just sufficient and the desired performance cannot be accomplished by the PID control process alone according as the performance requirement of a system becomes more strict. To keep robust performance of a system with respect to the unknown dynamics, it is necessary to reduce the system's sensitivity to the perturbation by applying an additional perturbation compensator as well as the nominal feedback controller. Taking a example, the IC chip mounter in the semiconductor industry and the CNC machining center are usually equipped with an additional function in the servo controller to compensate such as friction or backlash.

To get best control performance for a given dynamical system, the plant dynamics and the information on the reference trajectory or even the external disturbance could be positively involved in the control logic. For example, the computed torque method for robot manipulator [5] is a model-based control technique where the nonlinear part and the gravity in the manipulator dynamics are compensated using the model parameters, reference trajectory, and feedback signals. As another example, we can take the zero-phase-error-tracking controller (ZPETC) [92,94,95] which is a feedforward control method using the preview signal of reference trajectory. It requires the inverse model of the closed-loop dynamics and enables to increase the tracking performance by rejecting the dynamic lag (phase) between the reference command and the controlled output of the closed-loop system. In some cases, the controller can be synthesized so that it accommodates to the external disturbances. Such an example is the repetitive controller [92,93], which can compensate repeatable disturbances with known frequencies in terms of the internal model principle.

Like the above examples, the sophisticated controllers designed considering the dynamic model of the plant and the reference trajectory corresponding to the work schedule could realize extremal performance achievable for the given nominal plant, but the following problem is the *robustness* to the perturbation (i.e., plant uncertainty) which is not involved in the controller design procedure. When some reliable models for the nonlinear characteristic or the external disturbance of a plant are available, these effects could be readily compensated by including the models in the control algorithm. However, the problem is that most of the perturbation is not identifiable and although possible, it cannot be used in controller design when the equation describing the physics of the perturbation is too complex to implement. Basically, the uncertainty is not uncertainty any more if it can be identified in advance. Already we have numerous solutions by many pioneers for the robustness issue, specifically in linear systems. However, a number of them are meaningful just in mathematical sense and not practical in engineering respect as far as the robust control solution assumes a model of uncertainty itself.

Focusing on the implementation practicality rather than on the mathematical completeness, the disturbance observers [39,44,45,54,55,71,97] attracted great interest as a simple robust control approach profitable for mechanical systems in the last ten to fifteen years. As well, there exists diverse versions of the disturbance compensation method based on the identical idea with the disturbance observer. Although there are a few criticisms on the disturbance observer technique, for example, that it is no more than an integral control [63], it can be regarded, in author's mind, as a realistic approach to achieve robust performance in that it utilizes only the nominal plant model to estimate the unknowns of the plant.

**Friction Modeling and Compensation:**

A central topic in mechanical systems control is the friction modeling and compensation. In normal situation with no external disturbances, the stick/slip motion due to the nonlinear friction around the zero velocity is the major obstacle to achieve fine tracking performance in many systems with contact-type bearings. The nonlinear friction depends on various phenomena such as hysteresis and position-dependent or temperature-dependent random behaviors [3,4]. Also, the microscopic frictional behavior near the zero velocity is very difficult to identify and denotes quite different physics from the macroscopic behavior [35,76]. Actually, how to attack the frictional effect has been a hot issue in the control field. As a solution to the friction problem, roughly speaking, there have been three kinds of approaches. The first method is the sophisticated friction modeling and the compensation based on the identified model [15,28]. The second is the adaptive technique where a few parameters of the simple friction model are updated on-line [8,24]. And the third is to use a kind of perturbation observer, for example the disturbance observer (DOB) in [36,71]. Unlike the first and the second, the third can be regarded as an indirect friction compensation using the nominal plant model without a direct friction model. One of the above three approaches could be selectively applied depending on the plant characteristics or they can be used complementarily to each other.

Using the perturbation observer investigated in this book, the effort to identify the friction characteristics of a plant can be greatly relaxed. In fact, in the perturbation estimate which is the output of the perturbation observer, the friction component cannot be distinguished from the other components. However, the perturbation observer actually will work as a friction compensator in many mechanical systems where the friction is the dominant uncertainty.

## 1.3 Robust Motion Control with Perturbation Observer

There were a number of works regarding the robust motion control of mechanical systems such as feed drive systems, CNC machining centers, automatic assembly systems, robot manipulators, and other manufacturing systems. Among them, a class of controllers in [9,13,19,33,34,55,70,71,94,97,101,105] stresses robust performance in terms of the two-degree-of-freedom control structure shown in Fig. 1.1, which is composed of the feedback controller for nominal stability/performance and the additional perturbation observer to attenuate the effect due to the plant uncertainty.

In fact, the robust optimal control theories are much mathematical and they cannot be readily understood by industrial engineers while the perturbation observer is physically intuitive, structurally simple, and can be easily implemented. A remarkable feature of the perturbation observer is that it can *adaptively* estimate time-varying perturbations (i.e., plant uncertainties and external disturbances) and produce just minimal control effort to compensate the magnitude of the current perturbation. Hence, the conservativeness of linear robust controllers can be considerably relaxed if the perturbation observer is jointly used with them. Although there exist diverse versions related to perturbation compensation techniques, the term of *perturbation observer* (equivalently, perturbation compensator) will be used as a general terminology including those all in this book.

There exist a class of perturbation observers with common characteristics. First of all, a typical form of the perturbation observer is the disturbance observers (DOB) [39, 44, 45, 54, 71, 97], which has been formulated for linear systems in frequency domain with output feedback condition. Another notable form of the perturbation observer is the time delay estimation methods or the so-called time delay controller (TDC) [34, 37, 64, 66, 106], which directly considers general nonlinear plants in time domain with the assumption that the full state is available. The classical internal model control (IMC) theory [65] also belongs to this class, which has a duality relationship with DOB. In addition, there are the adaptive inverse controller [99], the adaptive robust controller(ARC) [101], the momentum feedback disturbance observer (MFDOB) [70], the model-based disturbance attenuator (MBDA) [13], and the robust internal-loop compensator (RIC) [41]. The above schemes can be classified into two sets again. The one is the inverse-plant-model-based compensators including DOB, IMC, and TDC. The other is a sort of reference model following internal feedback controllers, which includes such as ARC, MBDA, and RIC. The perturbation observers were widely applied to motion control systems due to the simplicity but the excellent performance. For example, the time-delayed controllers (TDC) were applied to robot manipulators [34, 66] and to magnetic suspension systems [37, 64]. The disturbance observers (DOB) were applied to robot manipulators [44, 45, 71] and to XY tables [39, 54]. As well, the adaptive robust controller(ARC) [101] and the model-based disturbance attenuator (MBDA) [13] were applied to CNC machining centers.

On the other hand, an unified view is possible for this class of perturbation observers. First, these schemes aim to achieve nominal plant dynamics in the perturbation compensation loops. Second, they use just the nominal plant model and the input/output signals of the plant. Third, the perturbation estimate by these methods corresponds to the deviation, which is converted into torques, of the actual plant behavior from the nominal dynamics. Although these are formulated in some different points of view, most of them finally must meet in similar forms in the discrete-time implementation.

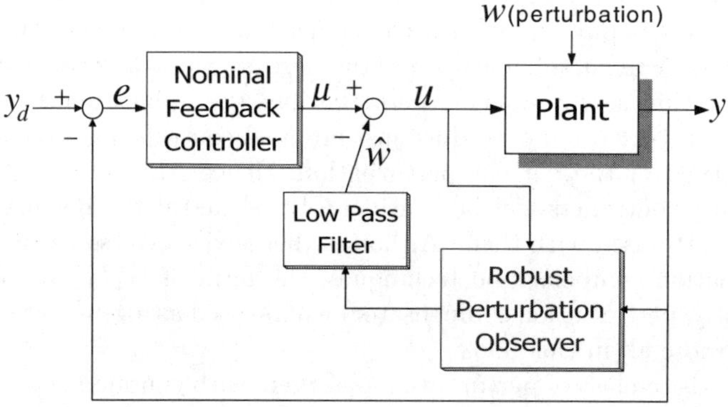


**Fig. 1.1.** Perturbation-observer-based robust motion control.

Another merit of the perturbation observer is that it can be designed in a decoupled manner independent of the main feedback controller and could be plugged in the existing control systems to increase the control performance without altering the main control architecture. As shown in the two-degree-of-freedom control system in Fig. 1.1, the perturbation observer functions as an *inner-loop regulator* which compensates all the perturbation ($w$) to the nominal plant. With the successful working of the perturbation observer, the inner-loop behaves almost like the nominal plant. The total control input in Fig. 1.1 is determined as the sum of the main tracking control input ($\mu$) and the perturbation compensation input($\hat{w}$), i.e., $u = \mu - \hat{w}$. As a consequence, the robust motion control performance can be achieved with the action of the perturbation compensation loop. Also, the perturbation observers can be used as a subsidiary function to improve the robustness property of traditional control theories. Taking some examples, first, several researchers proposed improved variable structure controllers using the time delay estimation technique to reduce the upper bound of discrete switching input [18,19,66,90]. Second, to construct a robust state estimator for nonlinear systems, Moura et al. [69] and Chang et al. [10] proposed the sliding perturbation observer (SPO) and the time delay observer (TDO), respectively, where a perturbation observer improves the robustness property of existing state observers.

In Chapters 2, 3, and 4, new points of view on the perturbation observer design and analysis are presented and improved perturbation compensation algorithms are suggested to overcome the performance limit of existing methods. The major motives are to utilize the preview signal of reference trajectory and introduce the notion of *residual perturbation* and *hierarchical perturbation compensation.*

## 1.4 Robust State Estimation with Perturbation Observer

A reliable state estimate of a process is indispensable not only for control purpose but also for other applications such as navigation of spacecraft, monitoring, and fault diagnosis in mechanical systems. Since the mathematical model used in the state estimator is only an approximation to the physical process and the actual plant is usually affected by external disturbances, there always exists an discrepancy between real states and estimates. However, the state estimators with the innovation process alone have a limit to successfully reconstruct the state of the plant with diverse uncertainties. As an example, conventional state estimators such as Luenberger observer or Kalman filter produce biased estimates for biased external disturbances.

In fact, the accuracy of state estimation largely depends on how well the physics of a real plant is reflected in the estimator structure. For robust performance of a state estimator, the real situation occurring in the plant such as the actuator saturation or other parasitic effects should be possibly copied to the state estimator structure so that the gap between the actual plant behavior and the nominal dynamics in the estimation algorithm should be maintained as small as possible. There exist some approaches for robust state estimation [10,27,69,84,98] and for robust Kalman filtering [42,75,80]. Most of these are equipped with an additional function to reflect the uncertainties as well as the conventional state observer or the conventional Kalman filter. Also, by introducing the perturbation observer of this book into the conventional state estimator, a robust state estimation performance could be achieved.

Hence, in Chapter 5, synthesized is the *combined observer* which is a combined setting of the linear state estimator and the discrete perturbation observer. The combined observer is not only a robust state estimator but also a output feedback perturbation observer, where the full-state reconstruction and perturbation observation are performed simultaneously. Moreover, the idea of combined observer is extended to stochastic systems and we reformulate the discrete Kalman filter with the perturbation observer. As a result, we have a modified recursive state estimation algorithm which includes the perturbation estimate update process.

**Feedback Connection of Feedback Controller, State Estimator, and Perturbation Observer:**

When the combined state estimator–perturbation observer is joined with the feedback controller, the overall observer-based control logic for robust motion control with robust state estimation can be completed. As a result, it takes the form of Fig. 1.2, which illustrates the input/output signal flow in the combined set of feedback controller, state estimator, and perturbation observer. In Fig. 1.2, the combined observer supplies state estimates ($\hat{x}$) to the feedback controller requiring full-state feedback and it also produces perturbation estimates ($\hat{w}$) which is applied as a supplementary control input to keep robust performance in the presence of plant uncertainties. As shown, both the state estimator and the perturbation observer commonly receive input ($u$) and output ($y$) of the plant to construct innovation processes and the two parts of the combined observer are dynamically coupled in such a way that the output of the one is needed as an input of the other and vise versa. These issues will be explained in Chapter 5.

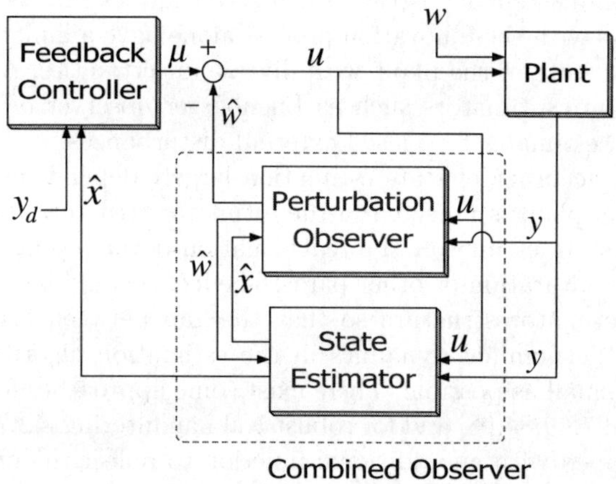


**Fig. 1.2.** Connection of feedback controller, state estimator, and perturbation observer.

## 1.5 Coarse/Fine Dual-Stage System

Two major sources of hindering control performance in systems with conventional actuators e.g., electrical motors or hydraulic actuators, may be the nonlinear friction in low speed motions and the resonance mode in high frequency motions. On the other hand, *fine actuators* such as piezoelectric actuators are free of friction and enable much higher resolution of motion although these are limited in travel range, at most, to a few hundreds of microns. The purpose of compound actuation using coarse/fine dual-stage is to realize a system with high precision positioning capability while keeping the travel range of conventional actuator. In the compound actuation systems, the coarse actuator and fine one are complementary to each other. That is, the coarse actuator offers large workspace and large actuation power while the fine actuator does high resolution of motion. There were several approaches where the concept of coarse/fine compound actuation is adopted, for example, macro/micro robot manipulators [30,40,78,82,83,104] and dual-stage XY positioners [49,50,53,87]. However, these days, the research on the coarse/fine dual-stage servo is most active in data storage systems [21,29,57,79,103], where the second actuator fabricated using the micro-machining technology enables high speed track following.

In Chapter 6, a control strategy for the coarse/fine dual-stage system will be investigated, where the perturbation compensator plays a crucial role to achieve robust tracking performance.

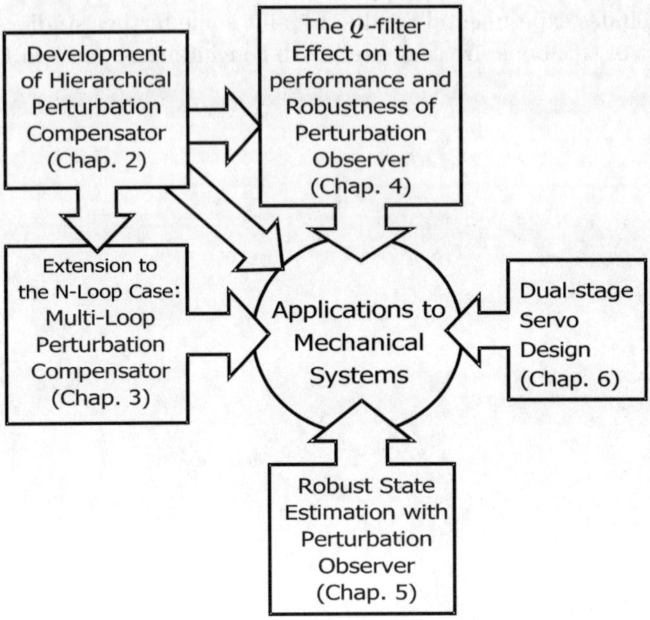


**Fig. 1.3.** Organization of contents.

## 1.6 Outline of the Book

The highlight of this book begins at Chapter 2 and ends at Chapter 6 and each chapter is described in a self-contained manner. Figure 1.3 shows the overall organization of contents. First, in Chapter 2, the hierarchical perturbation compensator (HPC) is proposed and a robust tracking controller which enables chattering-free smooth sliding mode is suggested. In Chapter 3, the notion of residual perturbation and the hierarchical perturbation compensation is extended to general $n$-loop case and the multiloop perturbation compensator (MPEC) is formulated. In Chapter 4, the perturbation compensation loop is analyzed in discrete-time viewpoint and it is clarified how the performance and stability of the perturbation observer is changed depending on the perturbation observer parameters. In Chapter 5, a robust state estimator is proposed by combining the conventional Luenberger observer and the Kalman filter with the perturbation observer. A recursive algorithm for simultaneous estimation of plant state and perturbation is given. In Chapter 6, a control architecture for the coarse/fine dual-stage system is constructed and it is applied to the micro-teleoperation. To support the validity of developed theories, each chapter includes experimental results. Finally, some further studies related to the subject of this book are suggested with concluding remarks in Chapter 7.

# 2

# Robust Tracking Control with Hierarchical Perturbation Compensation

The notion of perturbation observer is introduced and the idea of hierarchical compensation is suggested. This chapter will give the motive of applying perturbation observer technique for robust tracking control when a mechanical system is under heavy perturbation. In addition, we discuss a smooth sliding control in terms of perturbation observer. Regarding the stability analysis, the results in this chapter lead us to view the feedback loop from another angle.

## 2.1 Introduction

Although the optimal control methods such as $H_\infty$ control and LQG/LTR control achieve the robust stability for a specified uncertainty model, the algorithms inevitably require high deterministic gains to cover the worst case of the plant uncertainty. As a result, they would make the closed-loop system very conservative and give rise to an implementation issue. On the other hand, the class of motion controllers in [9,13,19,33,34,41,45,55,66,71,97,101,106] stresses the robust performance in terms of the two-degrees-of-freedom control structure composed of the nominal feedback controller and additional *robust compensator* to attenuate the perturbation to the plant. These have been successfully applied to mechanical systems such as CNC machine tools [13, 19, 101], robot manipulators [34,45,66], feed drive systems [41,55], and others. The class of the robust perturbation compensators is characterized by the *adaptive and integral property.* Unlike the deterministic high gain controls, it generates just minimal control effort to regulate the plant behavior to the nominal dynamics. In addition, it is physically intuitive and easy to implement.

Although mechanical systems have various nonlinear dynamic characteristics, the most significant parasitic effect hindering the control performance may be the nonlinear friction around the zero velocity. As a solution to the friction compensation, there have been three kinds of approaches. The first is the sophisticated friction modeling and compensation based on the identified model [15]. The second is the adaptive compensation technique where some parameters of a simple friction model are updated in real-time [8]. And the third is to use a perturbation compensator aforementioned, e.g., disturbance observers (DOB) in [36]. Unlike the first and the second, the third is an *indirect* friction compensation method without a direct friction model, where the nonlinear friction is considered as a component of the perturbation in addition to the other modeling errors and external disturbances. In this chapter, a hierarchical perturbation compensation method will be presented for advanced perturbation attenuation performance.

The sliding mode control (SMC) is very attractive for the tracking control of mechanical systems because of its clear definition of the closed-loop performance, the systematic design procedure using the sliding surface, and the robustness property to the system uncertainty. Another merit of the SMC is that it does not impose restriction on the type of the plant model since it is a time domain formulation, where linear and nonlinear systems are all admissible. The robustness of the conventional sliding control is achieved by the variable structure control (VSC), i.e., a discrete switching control while the switching gain is determined based on the Lyapunov asymptotic stability for the upper bound of the plant uncertainty. However, the limit of physical plants such as the bandwidth limit of actuator dynamics and the imperfection of sensory data inevitably produces the chattering problem, which has been a critical issue in the sliding mode control. To remedy this problem, several smoothing techniques for the sliding control were suggested [85,107]. However, the controller switching when entering or exiting the boundary layer [85] may result in large transients depending on the controller type. Also, the effect of the frequency shaping of the sliding surface [107] is very restrictive.

The robust tracking controller proposed in this chapter is a sort of chattering-free smooth sliding controller. It achieves robust performance by the perturbation compensator instead of the discrete switching control. There were a few works [18,19,66,69] closely related to the approach in this chapter, where a perturbation compensation scheme is adopted to enhance the robustness of the conventional sliding controller. However, the highlight of this chapter is the hierarchical perturbation compensator (HPC) which uses both feedforward and feedback signals.

This chapter is organized as follows. First of all, a nominal tracking controller is designed using the sliding surface in Section 2. In Section 3, some novel viewpoints on the perturbation compensator design are given, a hierarchical algorithm is developed, and the perturbation compensated sling mode control (PCSMC) is synthesized. Section 4 is devoted to the stability analysis. Experimental results are shown in Section 5. Finally, summary is followed in Section 6.

## 2.2 Tracking controller Design

Let's consider the robust tracking control structure in Fig. 2.1. As shown, it has a two-degrees-of-freedom structure composed of the nominal tracking controller and the perturbation compensator for robust performance, where the inner and outer loop controller can be designed independently. The total control input is determined as sum of the tracking control input($\mu$) and the perturbation compensation input($\hat{\psi}$) as

$$u(t) = \mu(t) - \hat{\psi}(t). \tag{2.1}$$

In this section, we design the nominal tracking controller using the sliding surface.

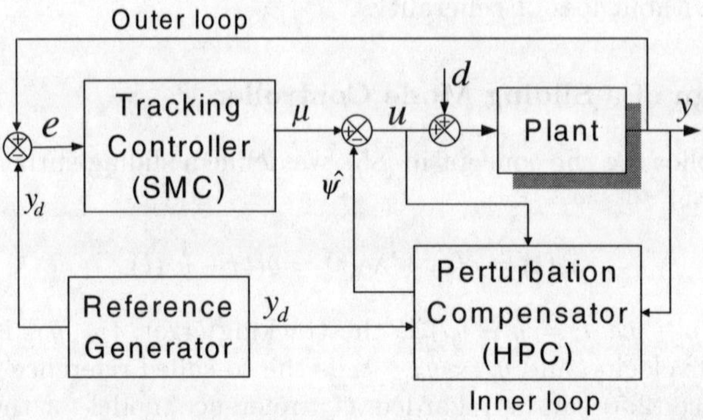


**Fig. 2.1.** Perturbation compensator based robust tracking control (PCSMC).

### 2.2.1 Nominal Model

A number of mechanical systems can be modeled as a Lagrange system as

$$H(y)\ddot{y}(t) + C(y,\dot{y}) + G(y) = u(t) + F_f(t) + h(t) + d(t) \tag{2.2}$$

where $H(y) \in \mathcal{R}^{n\times n}$ is the inertia matrix, $y \in \mathcal{R}^{n\times 1}$ is the generalized coordinate, $C(y,\dot{y}) \in \mathcal{R}^{n\times 1}$ is the inertial cross couplings, $G(y) \in \mathcal{R}^{n\times 1}$ is the gravitational force, $u(t) \in \mathcal{R}^{n\times 1}$ is the input vector, $F_f(t) \in \mathcal{R}^{n\times 1}$ is the friction force, $h(t) \in \mathcal{R}^{n\times 1}$ is the unmodeled dynamics such as the flexible mode, and $d(t) \in \mathcal{R}^{n\times 1}$ is the external disturbance. On the other hand, predominant behaviors of most mechanical systems can be described by the linear mass-damper model. Representing Eq. (2.2) in terms of the nominal parameters $(H_n, B_n)$, i.e, matrix of inertia and viscous damping coefficient, respectively, we have

$$H_n\ddot{y}(t) + B_n\dot{y}(t) = u(t) + \psi(t) \tag{2.3}$$

where $\psi(t)$ denote the *effective* perturbation which disturbs the nominal plant. It has the following components composed of model uncertainty and external disturbances:

$$\psi(t) = \Delta H\ddot{y} + f_c(t) + d(t) \tag{2.4}$$

where the second $f_c(t)$ includes all the other model uncertainties except for the first inertia uncertainty. To construct the proposed controller in Fig. 2.1, the inertia parameter $H_n$ for each motion axis is essential but the damping parameter $B_n$ is optional. To clearly describe the proposed controller, we consider the case that the model (2.3) is a decoupled MIMO system, i.e., the nominal parameters $(H_n, B_n)$ are diagonal matrices. So, in the rest of this chapter, we regard Eq. (2.3) as a SISO system corresponding to a local coordinate without loss of generality.

### 2.2.2 Design of a Sliding Mode Controller

Above all, following the concept in [85], we define a sliding surface, a dynamic hyperplane, as

$$s_e(t) = \dot{e}(t) + \lambda e(t) = \dot{y}(t) - \dot{y}_r(t) \tag{2.5}$$

where $(e = y - y_d,\ \dot{e} = \dot{y} - \dot{y}_d)$ is the tracking error, $(y_d, \dot{y}_d)$ is the desired position and velocity, and $\dot{y}_r = \dot{y}_d - \lambda e$ is the so-called reference velocity. The sliding surface (2.5) can be regarded as a reference model for the closed loop behavior and the positive constant $\lambda$ denotes the desired bandwidth.

First, an equivalent input with respect to the nominal model (2.3) is determined as

$$\dot{s}_e(t) = 0 \ \rightarrow \ u_{eq}(t) = H_n\ddot{y}_r(t) + B_n\dot{y}(t), \tag{2.6}$$

which generates a sliding mode if the initial condition is on the sliding surface and the plant dynamics is exactly the same as the nominal model. Second, a switching input is needed to generate the approaching mode to the sliding surface. In conventional sliding controllers, the switching input has the variable structure like $u_{sw} = -K\text{sgn}(s_e(t))$ where the gain $K$ satisfies the Lyapunov asymptotic stability for the upper bound of plant uncertainty. The discrete switching input makes the sliding surface an *invariant set* and so gives the robustness property to the system by keeping $\dot{s}_e(t) \approx 0$ in spite of the perturbation [85]. However, since the discrete switching of infinite bandwidth is physically impossible, the perfect sliding mode cannot be produced and the high switching gain would give rise to control chattering, which may excite unmodeled resonance modes. Hence, instead of the discrete switching input, we apply a *smooth* switching input such as $u_{sw} = -Ks_e(t)$. This input will produce low frequency switchings around the sliding surface.

Combining the equivalent input (2.6) with the smooth switching input, a tracking controller for nominal performance/stability can be determined as

$$\mu(t) = u_{eq}(t) + u_{sw}(t) = H_n \ddot{y}_r(t) + B_n \dot{y}(t) - K s_e(t) \tag{2.7}$$

which is a PD controller with feedforward terms. By applying the control input (2.7) to the plant (2.3), we have the closed-loop error dynamics as

$$H_n \dot{s}_e(t) + K s_e(t) = \psi(t). \tag{2.8}$$

As shown, the closed-loop system is persistently perturbed by the effective perturbation $\psi(t)$. Although there is no chattering problem, the robust sliding mode to the perturbation cannot be guaranteed by the input (2.7). To recover the robustness of the conventional sliding controller, the perturbation compensator is introduced in the next section.

## 2.3 A Class of Perturbation Observers

There were various formulations related to the perturbation observer (equivalently, compensator) with different names. First of all, the disturbance observer (DOB) [9, 45, 55, 71, 97] is very popular for robust motion controls in recent years. The DOB was formulated in frequency domain with respect to the linear system with output feedback condition. Another notable form of the perturbation observer is the so-called time delay controller (TDC) [33, 34, 66, 106]. It has been formulated directly for nonlinear plants in time domain under the assumption that the full state is given. The classical internal model controller (IMC) [65] also belongs to this class and it has a duality relationship with DOB. In addition, we can find some other forms of the perturbation observer such as the adaptive inverse controller [99], the adaptive robust controller(ARC) [101], the model-based disturbance attenuator (MBDA) [13], and the robust internal loop compensator (RIC) [41]. For the class of perturbation observers, a unified view is possible. That is, they give robustness to the system by achieving the nominal plant dynamics in the inner loop of a control system. Commonly, the observers estimate the perturbation to the plant by using the *indirect* perturbation estimation model composed of the nominal model and plant input/output signals. Although these are formulated in some different manners, the basic philosophy is the same.

At this point, it is necessary to figure out the concept of perturbation observers. Above all, from the plant model (2.3), the real perturbation in Eq. (2.4) can be *equivalently* expressed as

$$\psi_{eq}(t) = H_n \ddot{y}(t) + B_n \dot{y}(t) - u(t). \tag{2.9}$$

Then, the fundamental implementation form of all perturbation observers is simply

$$\hat{\psi}(t) = \psi_{eq}(t-L) = H_n\ddot{y}(t-L) + B_n\dot{y}(t-L) - u(t-L), \tag{2.10}$$

where $L$ is the control interval(or sampling time) in the digital controller. It is inevitable to adopt one step delayed signals to satisfy the causality between input and output samples. As shown in Eq. (2.10), the class of perturbation observers is based on the simple intuition that if all the states are available, the perturbation of the plant can be effectively estimated using the nominal model and one step delayed input/output signals. When only the position measure is available, a primitive way to obtain the other states is to take filtered derivatives for the position output. Another is to introduce a model-based state estimator.

In fact, Eq. (2.10) corresponds exactly to the time delay controller (TDC). Although it was not commented in the TDC formulations [33,34,66,106], the influence of sensor noise in Eq. (2.10) should be attenuated through a low pass filter for practical implementation of the perturbation observer. In that point, the disturbance observer (DOB) [45,55,97] or the internal model controller (IMC) [65] is more systematically designed by introducing a low pass filter. For example, for the linear plant, $y(s) = P(s)u(s)$ , the DOB has the form of

$$\hat{\psi}(s) = Q(s)\left(P^{-1}(s)y(s) - u(s)\right) \tag{2.11}$$

where the low pass filter $Q(s)$(so-called $Q$-filter) is closely related to the closed loop performance.

## 2.4 Three Points of View on the Perturbation Compensation

In this section, we present three kinds of viewpoint on the perturbation of a plant and develop the corresponding perturbation observers. First of all, substituting the total control input of Eq. (2.1) into the plant dynamics (2.3), we have the following relationship:

$$H_n\ddot{y}(t) + B_n\dot{y}(t) - \mu(t) = \psi(t) - \hat{\psi}(t) = \tilde{\psi}(t). \tag{2.12}$$

Here, we define a hyperplane describing the plant dynamics as

$$\varepsilon(\dot{y}, \ddot{y}, \mu, t) \triangleq H_n\ddot{y}(t) + B_n\dot{y}(t) - \mu(t). \tag{2.13}$$

This is equivalent to the perturbation compensation error in Eq. (2.12) and represents the deviation of the inner loop behavior from the nominal model. Then, keeping $\varepsilon(t) \approx 0$ means that the nominal dynamics is internally achieved. So, the perturbation compensator in Fig. 2.1 is expected to maintain $\varepsilon(t)$ as near zero as possible. Now, we present three possible points of view on the perturbation of a plant.

1) The first point of view corresponds to the expression (2.9), which can be regarded as *the perturbation w.r.t. the nominal dynamics of a plant.* All kinds of perturbation observers aforementioned are fundamentally based on this view. Using Eq. (2.1) and the definition (2.13), it can be rewritten as

$$\psi_n(t) = \psi_{eq}(t) = H_n\ddot{y}(t) + B_n\dot{y}(t) - u(t) = \hat{\psi}(t) + \varepsilon(\dot{y}, \ddot{y}, \mu, t). \quad (2.14)$$

2) Second, if we assume that the plant perfectly follows the desired trajectory, the real states in Eq. (2.9) can be replaced with the feedforward signals $(\dot{y}_d, \ddot{y}_d)$ of a reference trajectory, then we have *the perturbation w.r.t. the desired dynamics of a plant* as

$$\psi_d(t) = H_n\ddot{y}_d(t) + B_n\dot{y}_d(t) - u(t) = \hat{\psi}(t) + \varepsilon(\dot{y}_d, \ddot{y}_d, \mu, t). \quad (2.15)$$

3) Third, from the closed-loop error dynamics (2.31), we get another form of the equivalent expression for the perturbation as

$$\psi_s(t) = \hat{\psi}(t) + H_n\dot{s}_e(t) + Ks_e(t) \quad (2.16)$$

which can be interpreted as *the perturbation w.r.t. the closed loop sliding mode.*

In fact, the above three are equivalent expressions for the perturbation of a plant. However, the first and the second can be considered as the perturbation seen from the inner loop while the third is the one seen from the outer tracking control loop. Based on these three views, it is possible to construct corresponding perturbation observers.

### 2.4.1 Feedback Perturbation Observer (FBPO)

Following the basic concept in Eq. (2.10) using one sampling step delayed signals to implement a causal perturbation observer, the FBPO can be constructed based on the expression (2.14) or (2.9) as

$$\hat{\psi}_{FB}(t) = \psi_n(t - L) \quad (2.17a)$$

$$= \hat{\psi}_{FB}(t - L) + \varepsilon(\dot{y}, \ddot{y}, \mu, t - L). \quad (2.17b)$$

The FBPO estimates the perturbation *w.r.t.* the nominal dynamics of a plant and this compensation input will drive the inner loop in Fig. 2.1 to the nominal plant behavior under the tracking control input($\mu(t)$) as

$$H_n\ddot{y}(t) + B_n\dot{y}(t) \approx \mu(t), \quad \textit{i.e.,} \quad \varepsilon(\dot{y}, \ddot{y}, \mu, t) \approx 0. \quad (2.18)$$

The conceptual structure of the FBPO is denoted in Fig. 2.2, where the low pass filter (LPF)(i.e. $Q$-filter) is needed to attenuate the sensor noise(see Remark 2.2). The FBPO (2.17) is equivalent to the time delay controllers (TDC) [33, 34, 66, 106] or the time domain implementation form of disturbance observers (DOB) [55,71,97] without $Q$-filter.

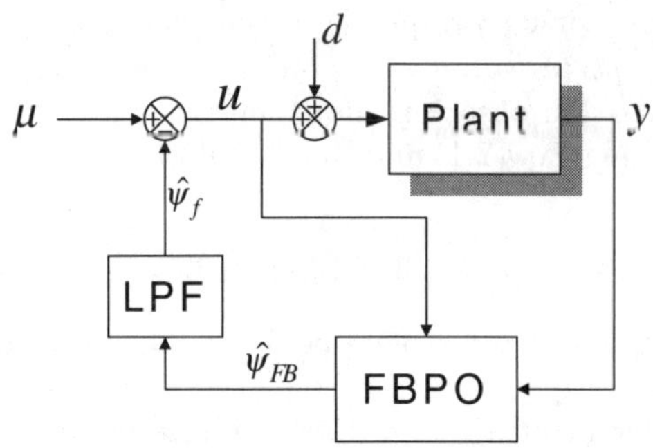


**Fig. 2.2.** FBPO loop.

### 2.4.2 Feedforward Perturbation Observer (FFPO)

In tracking control problems, the feedforward control technique using the reference trajectory is very useful as a subsidiary function to increase the control performance. Also, in the perturbation observer design, we can readily utilize the feedforward signals. The benefit of the reference trajectory is that it has no sensor noise and no dynamic lag between plant input and output unlike feedback signals. Based on the expression (2.15), we have the FFPO as

$$\hat{\psi}_{FF}(t) = \psi_d(t - L) \tag{2.19a}$$
$$= \hat{\psi}_{FF}(t - L) + \varepsilon(\dot{y}_d, \ddot{y}_d, \mu, t - L) \tag{2.19b}$$

where we are free from state estimation problem. The compensation input using the FFPO will make a plant approximately behave as the desired dynamics according to the reference trajectory as

$$H_n\ddot{y}_d(t) + B_n\dot{y}_d(t) \approx \mu(t), \quad i.e., \quad \varepsilon(\dot{y}_d, \ddot{y}_d, t) \approx 0, \tag{2.20}$$

while, the structure of the FFPO is denoted in Fig. 2.3.

### 2.4.3 Sliding Mode Perturbation Observer (SMPO)

Also, in terms of the equivalent (2.16) viewed from the closed loop sliding mode, the SMPO is given as

$$\hat{\psi}_{SM}(t) = \psi_s(t - L) \tag{2.21a}$$
$$= \hat{\psi}_{SM}(t - L) + H_n\dot{s}_e(t - L) + Ks_e(t - L) \tag{2.21b}$$

where the sliding surface trajectory is used. This formulation gives a different intuition on the perturbation compensation in that it is based on the closed loop error dynamics. However, it should be noted that the SMPO (2.21) can be applied only if the sliding mode controller in Eq. (2.7) is used as a nominal tracking controller in Fig. 2.1. In fact, as far as the input (2.7) is applied, the SMPO (2.21) is just another expression of the FBPO (2.17). The structure of the SMPO is denoted in Fig. 2.4.

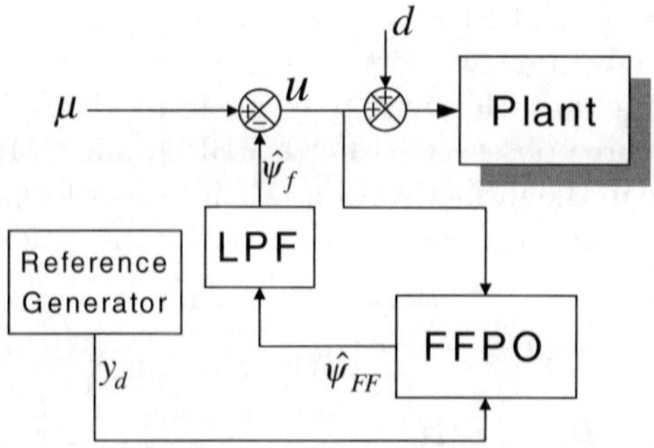


**Fig. 2.3.** FFPO loop.

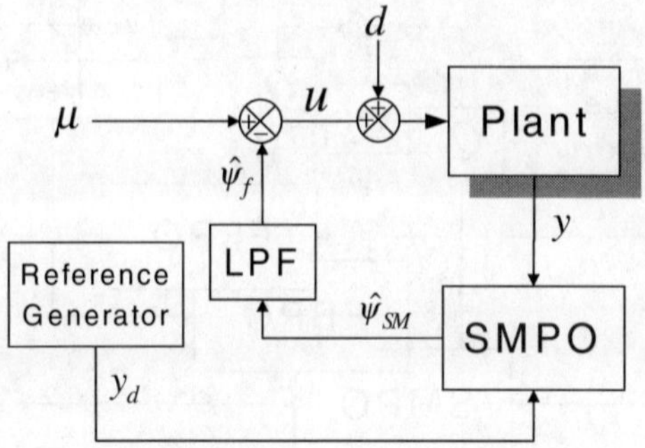


**Fig. 2.4.** SMPO loop.

## 2.5 Hierarchical Perturbation Compensator (HPC)

Since all the above three observers can effectively estimate the plant perturbation, if any one of them is combined with the tracking control input (2.7), the closed loop system will denote a robust sliding mode according to the error dynamics (2.31). However, a better perturbation attenuation performance is possible by combining the three observers. Considering the previous FBPO, its performance may be limited due to the dynamic lag between plant input/output and the sensor noise contained in the feedback signals. This problem can be complemented by FFPO using noise-free and dynamic lag-free feedforward signals. In addition, the property of SMPO which observes the perturbation in closed loop sliding mode can increase the overall performance.

At this point, we propose the *hierarchical perturbation compensator (HPC)* in Fig. 2.5, where the perturbation to the nominal plant will be attenuated hierarchically as follows. Above all, the FFPO($\hat{\psi}_1$) compensates a large amount of perturbation in a feedforward manner. The *residue* of the perturbation not rejected by the FFPO is compensated by the FBPO($\hat{\psi}_2$). The SMPO($\hat{\psi}_3$) once more compensates the residual perturbation still not observed by the combination of FFPO and FBPO. This approach makes it possible to overcome the performance limit of aforementioned perturbation observers which have *single-loop* structure. However, to compensate the residual perturbation by the HPC, the three observers, FFPO, FBPO, and SMPO should not be applied as they are in the former section. Their proper forms for the HPC will be followed. Prior to develop the HPC, we define the residual perturbation.

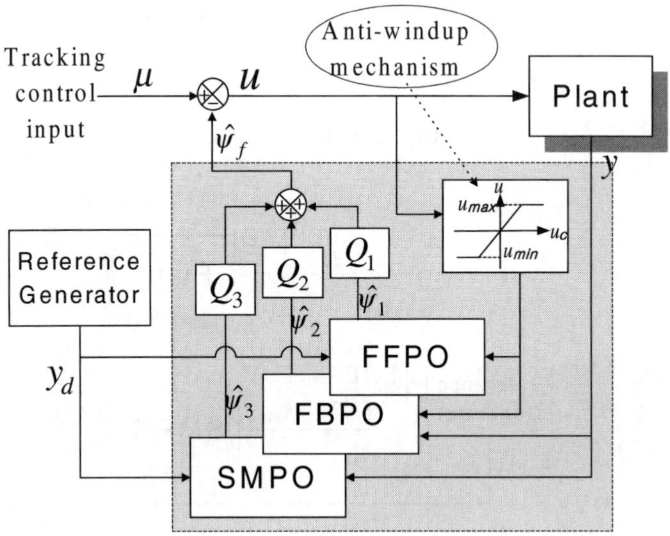


**Fig. 2.5.** Hierarchical perturbation compensator (HPC) with anti-windup mechanism.

**Definition 2.1 (Residual Perturbation).** When the FFPO ($\hat{\psi}_1$) of the HPC is applied, the residual perturbation, i.e., the perturbation compensation error is defined as

$$\tilde{\psi}_1(t) = \psi(t) - \hat{\psi}_1(t) \triangleq \psi_2(t). \tag{2.22}$$

The residual perturbation when applying the hierarchical sum of the FFPO ($\hat{\psi}_1$) and the FBPO ($\hat{\psi}_2$) of the HPC is

$$\tilde{\psi}_2(t) = \psi(t) - \hat{\psi}_1(t) - \hat{\psi}_2(t) = \psi_2(t) - \hat{\psi}_2(t) \triangleq \psi_3(t). \tag{2.23}$$

□

First, the FFPO in the HPC has the same form as the FFPO in Eq. (2.19) like

$$\begin{aligned}\hat{\psi}_1(t) = \psi_d(t-L) &= \hat{\psi}(t-L) + \varepsilon(\dot{y}_d, \ddot{y}_d, \mu, t-L) \\ &= H_n\ddot{y}_d(t-L) + B_n\dot{y}_d(t-L) - u(t-L). \end{aligned} \tag{2.24}$$

Replacing the real perturbation $\psi(t)$ in Eq. (2.22) with the equivalent $\psi_{eq}(t)$ in Eq. (2.14) and using the total compensation input $\hat{\psi} = \hat{\psi}_1 + \hat{\psi}_3 + \hat{\psi}_3$, the residual perturbation $\psi_2(t)$ due to the FFPO can be *equivalently* expressed as

$$\psi_{2eq}(t) = \psi_n(t) - \hat{\psi}_1(t) = \hat{\psi}_2(t) + \hat{\psi}_3(t) + \varepsilon(\dot{y}, \ddot{y}, \mu, t). \qquad (2.25)$$

Then, we have the FBPO for the HPC as

$$\hat{\psi}_2(t) = \psi_{2eq}(t - L) \qquad (2.26a)$$

$$= H_n\ddot{y}(t - L) + B_n\dot{y}(t - L) - u(t - L) - \hat{\psi}_1(t - L). \qquad (2.26b)$$

Similarly, replacing the real perturbation $\psi(t)$ in Eq. (2.23) with the equivalent perturbation $\psi_s(t)$ in Eq. (2.16), we have the equivalent for the residual perturbation $\psi_3(t)$ as

$$\psi_{3eq}(t) = \psi_s(t) - \hat{\psi}_1(t) - \hat{\psi}_2(t) = \hat{\psi}_3(t) + H_n\dot{s}_e(t) + Ks_e(t). \qquad (2.27)$$

Then, the SMPO for the HPC is given as

$$\hat{\psi}_3(t) = \psi_{3eq}(t - L) \qquad (2.28a)$$

$$= \hat{\psi}_3(t - L) + H_n\dot{s}_e(t - L) + Ks_e(t - L). \qquad (2.28b)$$

Finally, by combining the above three observers, (2.24), (2.26), and (2.28), the HPC input which utilizes both feedforward and feedback signals is determined as

$$\hat{\psi}_{HPC}(t) = \hat{\psi}_1(t) + \hat{\psi}_2(t) + \hat{\psi}_3(t) \qquad (2.29)$$

If any one or two loops of the HPC is cut in Fig. 2.5, the remaining loops will still function well as a perturbation compensator, but with reduced performance. Moreover, the sequence of the observers in the hierarchy of the HPC can be alternated with appropriate change of the definitions in Definition 2.1, but the effect of the hierarchical compensation still holds.

*Remark 2.2 (Noise Attenuation).* The FBPO (2.26) and the SMPO (2.28) respectively requires the full state including acceleration. Furthermore, in case only position measure is available, the velocity and the acceleration are usually determined through the numerical differentiation in which case the sensor noise may be greatly amplified. Also the FFPO (2.24) includes one-step delayed control input which already contains the sensor noise. Hence, for implementation of the HPC as well as FBPO, FFPO, and SMPO, a low pass filter (LPF) with *unity* DC gain, e.g., $Q(s) = a_c/(s + a_c)$ should be involved to attenuate the sensor noise as shown in Fig. 2.5. Then, the realizable form of (2.29) is

$$\hat{\psi}_{1f}(t) = \mathcal{Q}_1 \psi_d(t-L) \tag{2.30a}$$

$$\hat{\psi}_{2f}(t) = \mathcal{Q}_2 \left[\psi_n(t-L) - \hat{\psi}_{1f}(t-L)\right] \tag{2.30b}$$

$$\hat{\psi}_{3f}(t) = \mathcal{Q}_3 \left[\psi_s(t-L) - \hat{\psi}_{1f}(t-L) - \hat{\psi}_{2f}(t-L)\right] \tag{2.30c}$$

$$\rightarrow \hat{\psi}_{HPC}(t) = \hat{\psi}_{1f}(t) + \hat{\psi}_{2f}(t) + \hat{\psi}_{3f}(t) \tag{2.30d}$$

where $\mathcal{Q}_1, \mathcal{Q}_2, \mathcal{Q}_3$ are linear operators corresponding to the low pass filters, $Q_1, Q_2, Q_3$ in Fig. 2.5 □

*Remark 2.3 (Properties of Perturbation Observers).* Here, we remark common properties of the perturbation observers including FBPO, FFPO, SMPO, and HPC as well as the aforementioned class of perturbation observers.

1) (*Adaptive Control Property*) The perturbation observers, (2.17), (2.19), (2.21), and (2.29) have the adaptive control property. Actually, it generates as much control effort as is required to compensate the current perturbation. This is a distinguished characteristic from the other fixed gain robust control approaches where a sufficiently large deterministic gain is inevitable to cover the worst case of the perturbation. So, it can be called a signal-synthesis adaptive controller as in Hsia's comment for the time-delayed controller [33].
2) (*Integral Control Property*) As shown in the proposed perturbation observers (2.17), (2.19), (2.21), and (2.29), the current compensation input is determined by the one-step delayed input as well as the dynamic compensation terms. It characterizes the integral control property.
3) (*Repetitive Control Property*) Similarly to the time delay controller (TDC) in [106], the proposed perturbation observers can operate as a repetitive controller for a periodic disturbance if the delay time $L$ is matched to the disturbance period instead of the control interval.
4) (*Acceleration Control Property*) The perturbation estimation signal is physically equivalent to the acceleration. So, the perturbation observers change the inner loop of the two degrees of freedom control system in Fig. 2.1 to an open-loop acceleration control system as in Ohnishi *et al.*'s comment for disturbance observer (DOB) [71].
5) (*Feedback Linearization Property*) For nonlinear plants, if a linear system is chosen as the nominal model, which is the usual case in the perturbation observer design, the perturbation observers can be considered as a *signal-based* feedback linearizing controller since they reject all components of the perturbation including nonlinear terms. □

## 2.6 Perturbation Compensated Sliding Mode Control

By combining the nominal tracking controller (2.7) designed in the sliding control framework with the perturbation compensator, a robust tracking controller in Fig. 2.1, the perturbation compensated sliding mode controller (PCSMC) is synthesized. In the PCSMC, the robustness to the perturbation mostly depends on the perturbation compensator while the ground of the robustness in conventional sliding controllers is the discrete switching input. In fact, the perturbation compensator can be either FBPO, FFPO, SMPO or any hierarchical combination of them. The advantage of the HPC (2.29) in Fig. 2.5 is that it gives a chance to perform fine performance tuning through the subsidiary loops.

If the PCSMC input (2.1) including the perturbation compensation input is applied to the plant (2.3), the error dynamics (2.8) will be converted to

$$H_n \dot{s}_e(t) + K s_e(t) = \psi(t) - \hat{\psi}(t) = \tilde{\psi}(t). \tag{2.31}$$

Using the definition of Eq. (2.5) and $P = H_n^{-1}K$, it can be rewritten as

$$\ddot{e} + (P + \lambda)\dot{e} + P\lambda e = H_n^{-1}\tilde{\psi} \tag{2.32}$$

where the reduced perturbation from $\psi$ to $\tilde{\psi}$ directly means the enhanced robustness of the system and the increased tracking performance. If the norm of the perturbation compensation error $\|\tilde{\psi}\|$ is uniformly bounded in sufficiently small values, also the norm of the tracking error $\|e\|$ will be bounded by small values.

Here, we consider a Lyapunov function candidate, $V(t) = \frac{1}{2}s_e^T H_n s_e$. Taking the time derivative and substituting Eq. (2.31) into it, we have

$$\dot{V}(t) = s_e^T H_n \dot{s}_e = -s_e^T K s_e + s_e^T \tilde{\psi}. \tag{2.33}$$

If the perturbation compensation input maintains the compensation error $\tilde{\psi}$ uniformly bounded and the gain satisfies $K > \left|\tilde{\psi}(t)/s_e(t)\right|$ (except when $s_e = 0$), we get $\dot{V}(t) < 0$. Note that when $s_e = 0$, $V = \dot{V} = 0$ but $s_e = 0$ is an invariant set(i.e, equilibrium point). Then, according to the invariant set theorem [85], the tracking control system will be asymptotically stable with a smooth sliding mode. However, in practice, it is impossible to achieve the asymptotic stability in the presence of arbitrary perturbation. Instead, just the boundedness of the perturbation compensation error can be said, which will be addressed in the following stability analysis.

Taking the Laplace transform to Eq. (2.32), we have

$$\frac{e(s)}{\tilde{\psi}(s)} = \frac{H_n^{-1}}{s^2 + (P+\lambda)s + P\lambda} \triangleq G(s). \tag{2.34}$$

If the control gains are selected as $\lambda = P$, the error system (2.32) will be critically damped and the bandwidth of the closed loop motion becomes approximately $\omega_{BW} \approx \lambda = P$. In case of $\lambda = P$, Eq. (2.34) has the maximum at $s = 0$ in frequency response. So, the tracking error will be bounded in a finite time by

$$|e(t)| < \frac{H_n^{-1}}{\lambda^2} \max_t\{\left|\tilde{\psi}(t)\right|\}. \tag{2.35}$$

Then, it can be expected that *approximately the tracking error of the PCSMC is reduced in such a manner that if the gain is doubled, the error is reduced to one fourth.*

*Remark 2.4 (A Gain Tuning Rule).* The overall performance of the PCSMC in Fig. 2.1 is determined by the blending of the tracking control performance of the outer loop and the perturbation attenuation performance of the inner loop. While, the perturbation attenuation performance can be adjusted by increasing the cut-off frequency($a_c$) of the low pass filter in Eq. (2.30) to the *allowable maximum* considering the system uncertainty and hardware limitations. In fact, it is very difficult to say what is the gain set which produces the global minimum of the tracking error. It largely depends on the problem. However, if the prefiltered perturbation compensation signal $\hat{\psi}(t)$ in Eq. (2.30) and the driving signal $\tilde{\psi}(t)$ to the error dynamics in Eq. (2.32) have similar power spectrum, we can let the cut-off frequency of $Q(s)$ in Eq. (2.30) and that of $G(s)$ in Eq. (2.34) the same. Then, we can suggest a simple gain tuning rule such as $\lambda = P = a_c$. □

*Remark 2.5 (Feedforward Control Property).* If the perturbation is perfectly compensated, *i.e.*, $\tilde{\psi}(t) = 0$, the error dynamics (2.32) says that

$$\frac{y(s)}{y_d(s)} = \frac{s^2 + (P+\lambda)s + P\lambda}{s^2 + (P+\lambda)s + P\lambda} = 1 \tag{2.36}$$

in frequency domain. This means that the PCSMC law naturally includes the feedforward control property which compensates the dynamic lag between $(y_d,\ y)$ of the closed loop system. An issue in the feedforward controller design is the robustness with respect to the model uncertainty of the plant [94]. Since the proposed hierarchical perturbation compensator effectively rejects the plant uncertainty in the *acceleration* level, we can claim that the PCSMC has a robust feedforward control property. On the other hand, the plant dynamics under digital control depends on the sampling time and the unstable zeros may occur with fast samplings. In this case, it may be worthwhile to include a phase cancelling technique such as ZPETC [94] into the PCSMC structure. □

*Remark 2.6 (Direct Model-Based Compensation).* In the PCSMC controller, just the inertia parameter is required as a plant model(the viscous damping parameter is optional) and the other unknown dynamics is compensated *indirectly* by the perturbation compensator. However, if we have a reliable model which addresses some nonlinear dynamic behavior of the plant, a better control performance can be achieved by independently adding the model to the control input to *directly* compensate the dynamic effect. Usually, the nonlinear Coulomb friction of mechanical systems is identified as a function of velocity.

If we adopt, for example, a feedforward type friction compensator described as $\hat{F}_{nf}(t) = D_v \text{sgn}(\dot{y}_d)$ in [55], the total control input (2.1) is revised to $u(t) = \mu(t) - \hat{\psi}(t) - \hat{F}_{nf}(t)$. □

### 2.6.1 A Practical Consideration: Actuator Saturation

If there is an integral element in the controller, a windup phenomena may occur when the actuator is saturated but the control command in the control processor exceeds the actuator torque limit. The windup phenomena usually causes large transients but even destabilizes the system [23]. As stated in Remark 2.3, the perturbation observers have the integral control property. So, the actuator saturation must be considered as a practical problem. Specifically, when a time-optimal reference trajectory is applied, the actuator saturation is natural and the *anti-windup* is indispensable.

On the other hand, the basic idea of the anti-windup is to effectively break the integration loop during the saturation interval so that the control command is not over the actuator torque limit. Considering model-based state estimators such as Luenberger observer and Kalman filter, their performance depends on how much the observer reflects the real situation of the plant such as actuator saturation or other parasitic effects [25]. Since the actuator input predominantly determines the plant behavior over any other perturbations, it is essential to apply the same inputs both to the plant and to the observer so that the plant dynamics and the observer dynamics are *matched.* Since the perturbation observers are also based on the plant model, the same logic holds. As a matter of fact, if the saturation element representing the actuator power limit is included in the controller as an anti-windup mechanism and the limited control command is applied to the perturbation observer, the integral action is naturally stopped during the saturation interval.

Considering the HPC including FBPO, FFPO, and SMPO, a step delayed total input $u(t-L)$ enters the FBPO (2.26) and the FFPO (2.24). So, it should be limited by the anti-windup block of the saturation element as shown in Fig. 2.5. Then, the FBPO and FFPO signals would maintain proper values. As well, the SMPO (2.28) signal of the HPC can be a source of actuator saturation. So, if the control command one step before exceeds the limit torque(*i.e.*, $|u(t-L)| \geq u_{max}$), the SMPO output in Eq. (2.28) should be restricted as $\hat{\psi}_3(t) = \hat{\psi}_3(t-L)$ to stop the integral action.

### 2.6.2 Simulation

Figure 2.6 shows a simulation result for the XY positioner in Fig. 2.9. It denotes the radial tracking errors $e_r(t) = R - \sqrt{x^2(t) + y^2(t)}$ in tracking the circle (radius $R = 30mm$, tracking time $t_f = 2\ sec$) marked in the Fig. 2.9, while the trajectory of each axis for the circle is denoted in Fig. 2.7. While, the arbitrary large perturbation in Fig. 2.6(b) was added to both axes to

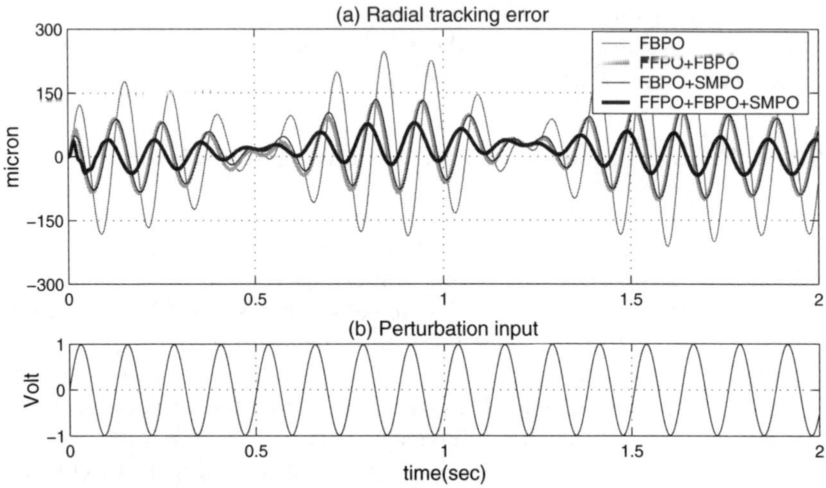


**Fig. 2.6.** (Circle($R = 30mm$) tracking simulation) (a) radial tracking errors, $e_r(t) = R - \sqrt{x^2(t) + y^2(t)}$, (b) sinusoidal disturbance input.

clearly show the hierarchical compensation effect. The PCSMC was applied with different perturbation compensators, in sequence, i) FBPO only, ii) the hierarchical sum of FFPO and FBPO, iii) the hierarchical sum of FBPO and SMPO, and iv) the hierarchical sum of FFPO, FBPO, and SMPO. Common to the cases, the tracking control gain in Eq. (2.7) was $\lambda = P = 100$ and the low pass filter for the perturbation compensator in Eq. (2.30) was $Q(s) = a_c/(s + a_c)$ with $a_c = 100$. As shown, the radial tracking error was gradually reduced as the perturbation observers are hierarchically added.

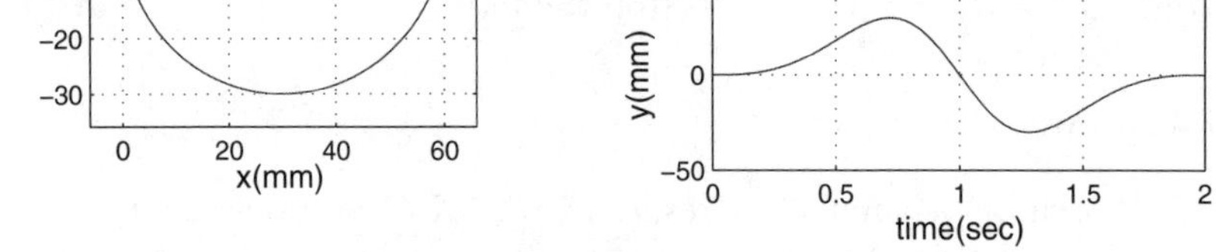


**Fig. 2.7.** Circle trajectory of radius $R = 30$ $mm$: $x(t) = R(1 - \cos\theta(t))$, $y(t) = R\sin\theta(t)$ with 2 $sec$ tracking time, where the rotation angle $\theta$ is given as a 5th order polynomial trajectory for the target angle, $\theta_f = 2\pi$.

## 2.7 Stability Analysis

In this section, we discuss the stability of the system when the proposed PCSMC is applied with the developed perturbation observers. In particular, we show that the hierarchical compensation law has an inherent property to increase the motion control performance. Above all, we have the following theorem.

**Theorem 2.7 (Input/Output Stability).** If the PCSMC controller (2.1) is applied to the plant (2.3), the tracking control loop is $\mathcal{L}_p$ stable($p \in [1, \infty)$) with respect to the pair $(\tilde{\psi}, s_e(t))$.

*Proof*: The closed loop error dynamics (2.31) can be rewritten as $\dot{s}_e + H_n^{-1}Ks_e = H_n^{-1}\tilde{\psi}$. Then, we can immediately see that the pair $(\tilde{\psi}, s_e)$ is a passive mapping. The $\mathcal{L}_p$ stability is valid for passive systems. A proof of the $\mathcal{L}_p$ stability for the passive systems can be found in [9]. (Q.E.D.)

Then, our attention is paid to the boundedness of the perturbation compensation error $\tilde{\psi}$ in Eq. (2.31). The fact that the effective perturbation in Eq. (2.4) is the function of state variables as well as external disturbance makes the stability analysis of the perturbation observer rather difficult. In this section, our primary concern is to discover under what conditions the perturbation compensation error $\tilde{\psi}(t)$ driving the error dynamics (2.31) converges to a small bound. If the boundedness of the compensation error is guaranteed, the sliding surface $s_e(t)$ and the tracking error $e(t)$ will be successively bounded according to Eqs. (2.31) and (2.5), respectively.

First, considering the effect of the FFPO, we expect that $\|\psi(t) - \hat{\psi}_{FF}(t)\| < \|\psi(t)\|$ when the FFPO (2.19) is applied. In fact, the inequality cannot be shown rigorously since the feedforward signal in $\hat{\psi}_{FF}$ has no relation to the plant outputs in $\psi(t)$. However, as far as the perturbation bandwidth are not so fast compared with the control bandwidth, it is certain that the feedforward dynamic compensation and the integral property of the FFPO will considerably reduce the tracking errors synergically with the nominal tracking controller. In succession, the stability of the feedback compensation loop will be dealt with in the next section.

*Remark 2.8 (Properties of Mechanical Systems).* In mechanical systems, since the relationship of actuator input to the velocity is a passive mapping, the velocity is always bounded and the acceleration is also bounded as far as the actuator input is bounded, *i.e.*, $\dot{y}$, $\ddot{y} \in \mathcal{L}_\infty$. Only the position output may not be bounded. Also, the friction force gives positive effect to the system stability. Cross couplings such as the Coriolis force and the centripetal force are artificial forces *w.r.t.* the rotating coordinate system and their magnitudes are always bounded as far as $\dot{y} \in \mathcal{L}_\infty$. The other unmodeled dynamics such as high frequency resonance modes exist far beyond the servo bandwidth. Then, all the terms of the perturbation in Eq. (2.4) except external disturbances can be considered to be bounded in magnitude. □

### 2.7.1 Boundedness of the Perturbation Compensation Error

In this section, we find out the robust stability conditions when the feedback compensation loops of the FBPO (2.17) and the HPC (2.29) are applied to the plant (2.3) with the tracking controller $\mu(t)$ in Eq. (2.7). The SMPO (2.21) is not dealt with since it is actually another expression of the FBPO (2.17) as far as the tracking controller is given by Eq. (2.7). The main points in the stability proof are to derive an error dynamics of the perturbation compensation loop and then determine a robust stability condition required for the boundedness of the compensation error.

**Assumption 2.1** To make the analysis easy, we restrict our attention to *mechanical systems* in Eq. (2.3) with single-input/single-output. Moreover, the following assumptions are necessary.

(A1) The nominal plant is time-invariant.

(A2) The full state, $y, \dot{y}, \ddot{y}$ are given at every sampling instant of the digital controller.

(A3) The equivalent of the perturbation, $\psi_{eq}(t)$ in Eq. (2.9) perfectly reconstructs the real perturbation, $\psi(t)$ in Eq. (2.4), i.e., $\psi_{eq}(t) = \psi(t)$ □

Above all, from Eq. (2.31) and $\dot{s}_e(k) = \ddot{y}(k) - \ddot{y}_r(k)$ in Eq. (2.5), the acceleration signal can be expressed in discrete form as

$$\ddot{y}(k) = H_n^{-1}\left(\tilde{\psi}(k) - Ks_e(k)\right) + \ddot{y}_r(k) \tag{2.37}$$

where $\ddot{y}_r(k) = \ddot{y}_d(k) - \lambda\dot{e}(k)$. From this relationship, it is expected that *the time-delayed acceleration feedback in the FBPO (2.17) will dynamically change the perturbation compensation error.* Substituting Eq. (2.37) into the perturbation (2.4), we have the following relationship.

$$\psi(k) = (1 - H_n^{-1}H)\,\tilde{\psi}(k) + \zeta(k) \tag{2.38}$$

where

$$\zeta(k) = \left(-1 + H_n^{-1}H\right)Ks_e(k) + (H_n - H)\ddot{y}_r(k) + f_c(k) + d(k). \tag{2.39}$$

In the first place, let's look into the single loop of the FBPO under digital control. According to Assumption 2.1, we can let $\psi_n(k) = \psi_{eq}(k) = \psi(k)$. Hence, the FBPO (2.17) can be written in discrete form as $\hat{\psi}_{FB}(k) = \psi_n(k-1) = \psi(k-1)$. As a $Q$-filter in (2.30), let's consider a 1st order filter, $Q_{10}(s) = \frac{1}{\tau s+1}$. Then, according to the *backward rule*, $s = (1 - z^{-1})/L$, it can be discretely approximated as $Q_{10}(z) = \frac{\gamma}{1+\beta z^{-1}}$, where

$$\beta = \frac{-\alpha}{1+\alpha} \text{ and } \gamma = \frac{1}{1+\alpha}, \quad (\alpha = \frac{\tau}{L}, \beta - \gamma = -1). \tag{2.40}$$

Hence, the perturbation compensation error due to the FBPO is equal to $\tilde{\psi}(k) = \psi(k) - Q(z)\hat{\psi}_{FB}(k) = \psi(k) - Q(z)\psi(k-1)$. By expanding $Q(z)$, we have $(1+\beta z^{-1})\tilde{\psi}(k) = (1+\beta z^{-1})\psi(k) - \gamma\psi(k-1)$. After the delay operation of $z^{-1}(\bullet)(k) = (\bullet)(k-1)$, we have $\tilde{\psi}(k) + \beta\tilde{\psi}(k-1) = \psi(k) + (\beta-\gamma)\psi(k-1) = \psi(k) - \psi(k-1)$. Finally, by substituting $\psi(k)$ in Eq. (2.38) into the above equation and arranging it properly, the following error dynamics can be derived.

$$\tilde{\psi}(k) = \left(1 - \gamma H_n H^{-1}\right)\tilde{\psi}(k-1) + \eta_{FB}(k) \tag{2.41}$$

where $\eta_{FB}(k)$ is given as the variation of the $\zeta(k)$ in Eq. (2.39) during a sampling interval($L$) as follows.

$$\eta_{FB}(k) = H_n H^{-1}\left(\zeta(k) - \zeta(k-1)\right) = H_n H^{-1}\delta\zeta(k) \tag{2.42}$$

**Theorem 2.9 (Boundedness of the Compensation Error Due to the FBPO).** Let the FBPO (2.17) with the 1st order filter $Q_{10}(z) = \frac{\gamma}{1+\beta z^{-1}}$ with the coefficients in Eq. (2.40) be applied to the PCSMC in Fig. 2.1. Then, the perturbation compensation error converges to a small bound if the following conditions hold: *(i) variation of the external disturbance during a sampling interval(L) is bounded, and (ii) nominal inertia($H_n$) satisfies the following condition($\gamma$ is given in (2.40)),*

$$0 < H_n < \frac{2}{\gamma}H. \tag{2.43}$$

*Proof.* From the discrete polynomial (2.41), the boundedness of $\tilde{\psi}(k)$ for the bounded $\eta_{FB}(k)$ is guaranteed as far as $\|1 - \gamma H_n H^{-1}\| < 1$, i.e., $0 < H_n < (2/\gamma)H$ in Eq. (2.43). Next, we consider the boundedness of the driving function $\eta_{FB}(k)$. From Eqs. (2.39) and (2.42), $\eta_{FB}(k)$ will be bounded if all the variations, $\delta s_e(k)$, $\delta\ddot{y}_r(k)$, $\delta f_c(k)$, and $\delta d(k)$ are bounded. From Eq. (2.5), $\delta s_e(k) = \delta\dot{y}(k) + \lambda\delta y(k) - (\delta\dot{y}_d(k) + \lambda\delta y_d(k))$ and $\delta\ddot{y}_r(k) = \delta\ddot{y}_d(k) + \lambda\delta\ddot{y}_d(k) - \lambda\delta\dot{y}(k)$. Above all, the desired trajectory is $y_d(k)$, $\dot{y}_d(k)$, and $\ddot{y}_d(k) \in \mathcal{L}_\infty$. In terms of Remark 2.8 for mechanical systems, the velocity is $\dot{y}(k) \in \mathcal{L}_\infty$(in fact, $\dot{y}(k) \in \mathcal{L}_\infty \Leftrightarrow \delta y(k) \in \mathcal{L}_\infty$), the unmodeled term is $f_c(k) \in \mathcal{L}_\infty$, and then the variations are also $\delta\dot{y}(k) \in \mathcal{L}_\infty$, $\delta f_c(k) \in \mathcal{L}_\infty$. Hence, $\delta s_e(k)$, $\delta\ddot{y}_r(k)$, and $\delta f_c(k) \in \mathcal{L}_\infty$. Consequently, if the *variation* of the external disturbance during a sampling interval is bounded, i.e., $\delta d(k) \in \mathcal{L}_\infty$, we have $\eta_{FB}(k) \in \mathcal{L}_\infty$. (Q.E.D.)

In the second place, we investigate the hierarchical loop of the HPC. According to Assumption 2.1, the equivalents in Eqs. (2.25) and (2.27) are equal to the actual residual perturbations in Eqs. (2.22) and (2.23), respectively, i.e., $\psi_{2eq}(t) = \psi_2(t)$ and $\psi_{3eq}(t) = \psi_3(t)$. Then, the FBPO in the HPC (2.26) is equal to $\hat{\psi}_2(k) = \psi_{2eq}(k-1) = \psi_2(k-1) = \psi(k-1) - \hat{\psi}_1(k-1) = \psi(k-1) - \hat{\psi}_{FF}(k-1)$(in the HPC, $\hat{\psi}_1(k) = \hat{\psi}_{FF}(k)$). Similarly, the SMPO in the HPC (2.28) is equal to $\hat{\psi}_3(k) = \psi_{3eq}(k-1) = \psi_3(k-1) = \psi(k-1) - \hat{\psi}_1(k-1) - \hat{\psi}_2(k-1) = \psi(k-1) - \psi(k-2) - \delta\hat{\psi}_{FF}(k-1)$. Hence, the HPC

(2.29) can be written in discrete form as $\hat{\psi}_{HPC}(k) = \hat{\psi}_1(k) + \hat{\psi}_2(k) + \hat{\psi}_3(k) = 2\psi(k-1) - \psi(k-2) + \delta\hat{\psi}_{FF}(k) - \delta\hat{\psi}_{FF}(k-1)$. When the HPC is applied with the same $Q$-filter in Eq. (2.40), the perturbation compensation error is as follows.

$$\tilde{\psi}(k) = \psi(k) - Q(z)\hat{\psi}_{HPC}(k) \tag{2.44a}$$
$$= \psi(k) - Q(z)\left(2\psi(k-1) + \psi(k-2) - \delta^2\hat{\psi}_{FF}(k)\right) \tag{2.44b}$$

where $\delta^2(\bullet)(k) \triangleq \delta(\bullet)(k) - \delta(\bullet)(k-1) = (\bullet)(k) - 2(\bullet)(k-1) + (\bullet)(k-2)$. Following the same procedure as in the above FBPO case to derive the Eq. (2.41), we have the following error dynamics for the HPC case.

$$\tilde{\psi}(k) = \left(1 + \gamma - 2\gamma H_n H^{-1}\right)\tilde{\psi}(k-1) \\ - \left(1 - H_n H^{-1}\right)\gamma\tilde{\psi}(k-2) + \eta_{HPC}(k) \tag{2.45}$$

where $\eta_{HPC}(k)$ is given as

$$\eta_{HPC}(k) = \eta_{FB}(k) - \gamma\eta_{FB}(k-1) - H_n H^{-1}\gamma\delta^2\hat{\psi}_{FF}(k) \\ = H_n H^{-1}\left(\zeta(k) - (1+\gamma)\zeta(k-1) + \gamma\zeta(k-2) - \gamma\delta^2\hat{\psi}_{FF}(k)\right). \tag{2.46}$$

As shown, it is the function of $\eta_{FB}(k)$ in Eq. (2.42) and $\zeta(k)$ in Eq. (2.39).

**Theorem 2.10 (Boundedness of the Compensation Error Due to the HPC).** Let the HPC (2.29) with the 1st order filter $Q_{10}(z) = \frac{\gamma}{1+\beta z^{-1}}$ with the coefficients in Eq. (2.40) be applied to the PCSMC in Fig. 2.1. Then, the perturbation compensation error converges to a small bound if the following conditions hold: *(i) double variation(variation of the variation) of the external disturbance during a control interval(L) is bounded, and (ii) the nominal inertia $H_n$ meets the following condition ($\gamma$ is given in (2.40)),*

$$0 < H_n < \left(\frac{1+\gamma}{3\gamma}\right)2H. \tag{2.47}$$

*Proof.* The Eq. (2.45) can be rewritten as

$$\begin{bmatrix}\tilde{\psi}(k-1)\\ \tilde{\psi}(k)\end{bmatrix} = \underbrace{\begin{bmatrix}0 & 1\\ -(1-H_nH^{-1})\gamma & 1+\gamma-2\gamma H_nH^{-1}\end{bmatrix}}_{\triangleq A}\begin{bmatrix}\tilde{\psi}(k-2)\\ \tilde{\psi}(k-1)\end{bmatrix} + \begin{bmatrix}0\\ \eta_{HPC}(k)\end{bmatrix}. \tag{2.48}$$

Then, if the absolute values of the eigenvalues of the matrix $A$ are less than 1, the boundedness of $\tilde{\psi}(k)$ for the bounded $\eta_{HPC}(k)$ is guaranteed. By letting the larger eigenvalue $\lambda_1 < 1$ and smaller one $\lambda_2 > -1$, the stability condition (2.47) can be readily derived. In fact, we have the same result by using the *Jury's test* [23] for the discrete polynomial (2.45). Next, we consider the boundedness of the driving function $\eta_{HPC}(k)$ in Eq. (2.46). As shown, $\eta_{HPC}(k) \in \mathcal{L}_\infty$ as far as $\eta_{FB}(k) \in \mathcal{L}_\infty$ and $\delta^2\hat{\psi}_{FF}(k) \in \mathcal{L}_\infty$. From Eq. (2.19), $\delta\hat{\psi}_{FF}(k) \in \mathcal{L}_\infty$ as far as $\delta\mu(k) \in \mathcal{L}_\infty$. From Eq. (2.7), $\delta\mu(k) \in \mathcal{L}_\infty$ as far as $\delta s_e(k), \delta\ddot{y}_r(k) \in \mathcal{L}_\infty$. From the proof of Theorem 2.9, $\delta s_e(k), \delta\ddot{y}_r(k) \in \mathcal{L}_\infty$ for mechanical systems. So, $\delta\hat{\psi}_{FF}(k) \in \mathcal{L}_\infty$ and also its variation, $\delta^2\hat{\psi}_{FF}(k) \in \mathcal{L}_\infty$. Also, from the proof of Theorem 2.9, $\eta_{FB}(k) \in \mathcal{L}_\infty$ as far as $\delta d(k) \in \mathcal{L}_\infty$. Hence, according to Eq. (2.46), the function $\eta_{HPC}(k)$ will be bounded if $\delta^2 d(k) \in \mathcal{L}_\infty$, i.e., the variation of $\delta d(k)$(*double variation* of $d(k)$) during a sampling time is bounded. (Q.E.D.)

### 2.7.2 Effects of the Hierarchical Perturbation Compensation

By applying the HPC, we expect that the norm of the residual perturbation is gradually reduced as $\|\psi\| > \|\tilde{\psi}_1\| > \|\tilde{\psi}_2\| > \|\tilde{\psi}_3\|$. For example, if $\hat{\psi}_{FF} = 0$ and $\gamma = 1$(no filtering case, $\tau = 0$) in Eq. (2.46), we have $\eta_{HPC} = \delta\eta_{FB}$. So, the compensation error in Eq. (2.45) will be so much decreased. However, this performance enhancement would be possible under some constraints such as control bandwidth and the cost. Based on the results in the former section, we discuss the advantage of the hierarchical perturbation compensation and the corresponding issues.

First of all, from Eqs. (2.41) and (2.42) for the FBPO, we have the following transfer function.

$$\frac{\tilde{\psi}(z)}{\zeta(z)} = \frac{H_n H^{-1}\left(1 - z^{-1}\right)}{1 - \left(1 - \gamma H_n H^{-1}\right) z^{-1}} \triangleq T_{FB}(z). \tag{2.49}$$

As shown in the former simulation and the following experimental results, the feedforward compensation loop in the HPC obviously enhances the control performance. However, it is difficult to mathematically prove how it contributes to reduce the residual perturbation in Eq. (2.45). So, we regard the case when the feedforward loop is cut in the HPC, i.e., $\hat{\psi}_{FF}(k) = 0$ and compare it with the single loop case. Then, from Eqs. (2.45) and (2.46) for the HPC, the transfer function from $\zeta(k)$ in Eq. (2.39) to the residual perturbation is given as follows.

$$\frac{\tilde{\psi}(z)}{\zeta(z)} = \frac{H_n H^{-1}\left(1 - (1+\gamma)z^{-1} + \gamma z^{-2}\right)}{1 - \left(1 + \gamma - 2\gamma H_n H^{-1}\right) z^{-1} + (1 - H_n H^{-1})\gamma z^{-2}} \triangleq T_{HPC}(z). \tag{2.50}$$

On the other hand, in some articles related to the disturbance observer (DOB) [45, 55], the so-called $Q_{31}$-filter, $Q_{31}(s) = \frac{3\tau s+1}{(\tau s+1)^3}$ was popularly used. By the backward approximation, it can be discretized as

$$Q_{31}(z) = \frac{\gamma_1 + \gamma_2 z^{-1}}{1 + \beta_1 z^{-1} + \beta_2 z^{-2} + \beta_3 z^{-3}} \tag{2.51}$$

where the coefficients are $(\alpha = \tau/L)$

$$\beta_1 = \frac{-3\alpha}{1+\alpha}, \beta_2 = \frac{3\alpha^2}{(1+\alpha)^2}, \beta_3 = \frac{-3\alpha^3}{(1+\alpha)^3}, \gamma_1 = \frac{1+3\alpha}{(1+\alpha)^3}, \gamma_2 = \frac{-3\alpha}{(1+\alpha)^3}. \quad (2.52)$$

Following the above procedure to produce Eqs. (2.49) and (2.50) for the $Q_{10}$-filter, it is not so difficult to derive the compensation error dynamics in case of applying the $Q_{31}(z)$ as a low pass filter in the perturbation observers. As a result, we have the following transfer function

$$\frac{\tilde{\psi}(z)}{\zeta(z)} = \frac{H_n H^{-1}\left(1 + d_1 z^{-1} + d_2 z^{-2} + d_3 z^{-3}\right)}{1 + c_1 z^{-1} + c_2 z^{-2} + c_3 z^{-3}}, \quad (2.53)$$

where the coefficients for the FBPO case are

$$c_r = \beta_r - \gamma_r(1 - H_n H^{-1}) \text{ and } d_r = \beta_r - \gamma_r, \quad (r = 1, 2, 3) \quad (2.54)$$

and those for the HPC case are

$$\begin{aligned} &c_1 = \beta_1 - 2\gamma_1(1 - H_n H^{-1}), c_2 = \beta_2 - (2\gamma_2 - \gamma_1)(1 - H_n H^{-1}), \\ &c_3 = \beta_3 + \gamma_2(1 - H_n H^{-1}), \\ &d_1 = \beta_1 - 2\gamma_1, d_2 = \beta_2 - 2\gamma_2 + \gamma_1, d_3 = \beta_3 + \gamma_2. \end{aligned} \quad (2.55)$$

*Remark 2.11 (Performance Enhancement).* In Fig. 2.8, the performance of the HPC is compared with the FBPO through the frequency responses of Eqs. (2.49), (2.50), and (2.53), where no compensation case($\hat{\psi} = 0$) denotes $\psi(k) = H_n H^{-1}\zeta(k)$ from Eq. (2.38). Recall that the FBPO fundamentally represents existing class of perturbation observers. As shown, it is clear that the HPC so much enhances the perturbation attenuation performance for the same low pass filter and the $Q_{31}$-filter turned out to be more effective in performance. The frequency response actually depends on the value of $\alpha = \tau/L$. For fixed sampling time($L$), as the filter cut-off frequency gets larger(i.e., $\tau \to 0$ or $\gamma \to 1$), the slope of the graphs in Fig. 2.8 becomes stiffer and the gaps between the FBPO and the HPC gets wider for the same low pass filter. □

*Remark 2.12 (Robust Stability Margin).* However, from the robust stability conditions (2.43) and (2.47) on the inertia uncertainty, the stability margin approaches the infimums as $\gamma \to 1$. So the filter cut-off frequency should be properly limited. Comparing Eqs. (2.43) and (2.47) derived for $Q_{10}$-filter, the upper bounded of the inertia uncertainty is smaller in the HPC case than in the FBPO case. Hence, we can see that *the HPC reduces the robust stability margin at the cost of performance enhancement.* In Eq. (2.43) for the FBPO, the upper bound of the inertia has $2H$ in no filtering case(i.e, $\gamma = 1$). Some related results to this condition can be found in time delay controllers (TDC) [34, 106], where the filtering effect was not considered. □

*Remark 2.13 (Relaxed Disturbance Condition).* For the perturbation in Eq. (2.4) to be bounded, the external disturbance should be bounded($d(k) \in \mathcal{L}_\infty$). However, as shown in Theorem 2.9, the residual perturbation when the FBPO is applied is bounded as far as the variation of the disturbance during a sampling time is bounded($\delta d(k) \in \mathcal{L}_\infty$). Also, in Theorem 2.10, the residual perturbation when the HPC is applied is bounded as far as the *double* variation of the disturbance is bounded($\delta^2 d(k) \in \mathcal{L}_\infty$). This means that perturbation observers greatly relax the external disturbance condition for system stability. For example, if the external disturbance $d(t)$ is a ramp function, $\delta d(k)$ is constant, and $\delta^2 d(k)$ becomes zero. □

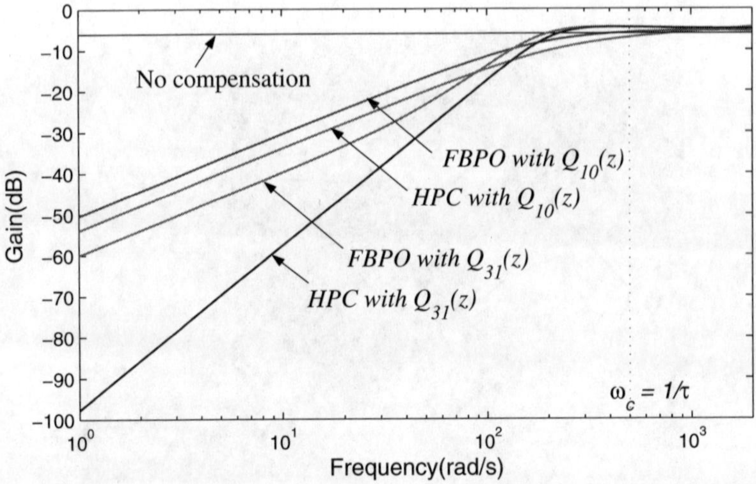


**Fig. 2.8.** Frequency response of the transfer function of the perturbation compensation error(when the control interval is $L = 1$ *msec*, the filter cut-off frequency is $a_c = 1/\tau = 500$, and the modeling error is $H_n = 0.5H$); (i) FBPO with $Q_{10}(z)$ in Eq. (2.49), (ii) HPC with $Q_{10}(z)$ in Eq. (2.50), (iii) FBPO with $Q_{31}(z)$ in Eqs. (2.53) and (2.54), (iv) HPC with $Q_{31}(z)$ in Eqs. (2.53) and (2.55).

## 2.8 Experiments

The effectiveness of the HPC was verified through the experiments for the XY positioner in Fig. 2.9. It is a ball-screw driven type where the travel range is 200 $mm \times 100$ $mm$, the screw lead is 5 $mm/rev$, and the rate speed is 250 $mm/sec$ at 3000 RPM. The closed loop control configuration is composed of the Pentium PC processor, D/A board(12 bit), BLDC motor with encoder(2500 $pulse/rev$), and counter board. Control commands were transferred to the motor driver at every $L = 1$ $msec$ using the PC interrupt under Windows 98 environment. The motor driver gain was fixed to generate the maximum torque($\pm 1.91$ $Nm$) at $\pm 5$ $V$ command. The parameters of the nominal model in Eq. (2.3) was experimentally identified under $2kg$ load condition as: $H_n[V/(m/s^2)] = (0.2020,\ 0.1665)$, $B_n[V/(m/s)] = (2.25,\ 1.35)$ for $(x, y)$ axis, respectively. As well, the Coulomb friction in the $(x, y)$ axis is, on the average, $C_{pos}[V] = (0.11,\ 0.09)$ in the positive direction and $C_{neg}[V] = (-0.13,\ -0.29)$ in the negative direction. In fact, the viscous damping term $B_n\dot{y}$ can be removed from the nominal model (2.3) since the linear damping force can be easily adjusted by the feedback control gain in Eq. (2.7). In this case, the model parameter required in the PCSMC controller is just the inertia parameter $H_n$ for each motion axis.

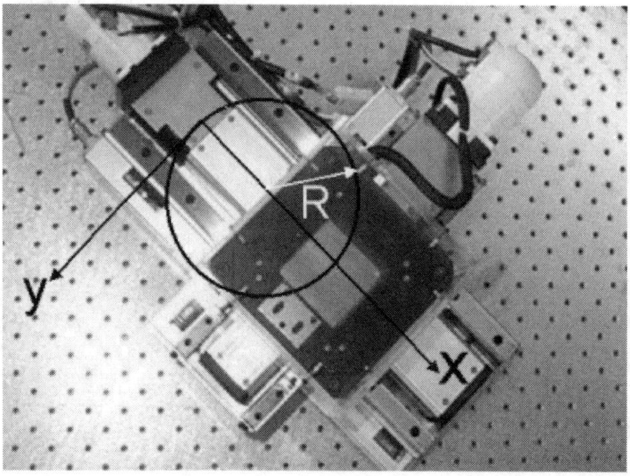


**Fig. 2.9.** Ball-screw driven XY positioner and the circle trajectory where the trajectory of each axis is given as $x(t) = R(1 - \cos\theta(t))$, $y(t) = R\sin\theta(t)$, the tracking time is $t_f = 2sec$, and the rotation angle $\theta(t)$ is determined as a 5th order polynomial for the target angle $\theta(t_f) = 2\pi$.

In all experiments, the velocity and acceleration required in the perturbation compensator were determined through the backward differentiation for encoder outputs and the data were filtered through the low pass filter, $1000/(s + 1000)$. The five combinations of the perturbation observers in Table 2.1 were compared, each of which corresponds to a special case of the HPC in Eq. (2.29). As a low pass filter, the 1st order filter $Q_{10}(z) = \frac{\gamma}{1+\beta z^{-1}}$ with coefficients in Eq. (2.40) was used.

**Definition 2.14 (Radial tracking error).** In the contour following problem using two-axis motions, the primary concern is the contour error rather than the tracking error in each axis. When the contour is a circle with radius $R$, the contour error corresponds to the the radial tracking error, $e_r(t) = R - r(t)$ with $r(t) = \sqrt{x^2(t) + y^2(t)}$. □

**Table 2.1.** Combinations of the perturbation observers.

| | |
|---|---|
| Case 1: FFPO | $\hat{\psi} = \hat{\psi}_1,\ \hat{\psi}_2 = \hat{\psi}_3 = 0$ |
| Case 2: FBPO | $\hat{\psi} = \hat{\psi}_2,\ \hat{\psi}_1 = \hat{\psi}_3 = 0$ |
| Case 3: FFPO+FBPO | $\hat{\psi} = \hat{\psi}_1 + \hat{\psi}_2,\ \hat{\psi}_3 = 0$ |
| Case 4: FBPO+SMPO | $\hat{\psi} = \hat{\psi}_2 + \hat{\psi}_3,\ \hat{\psi}_1 = 0$ |
| Case 5: FFPO+FBPO+SMPO | $\hat{\psi} = \hat{\psi}_1 + \hat{\psi}_2 + \hat{\psi}_3$ |

**Definition 2.15 ($L_2$ norm of tracking errors).** In addition to the peak values, a performance measure is needed to evaluate the degree of dispersion of the tracking errors. So, the $L_2$ norm of the radial tracking error is defined as $L_2[e_r] = \sqrt{(1/t_f)\int_0^{t_f} |e_r(t)|^2 dt}$. □

To clearly show the effect of the HPC, the arbitrary large sinusoidal perturbation shown in Fig. 2.13(b)(the peak is 30% of the maximum torque of the motor) was added to the control commands of both axes. First, Fig. 2.10 shows the contour errors in tracking the circle of radius $R = 0.5\ mm$ and tracking time $t_f = 2\ sec$ while the components of control input are shown in Fig. 2.11 and 2.12. In contour following problems using two-axis motions, a primary concern is the contour error rather than the tracking error in each axis. When the contour is a circle with radius $R$, the contour error corresponds to the the radial tracking error, $e_r(t) = R - r(t)$ with $r(t) = \sqrt{x^2(t) + y^2(t)}$. Second, Fig. 2.13 denotes the radial tracking errors for the circle trajectory with radius $R = 30\ mm$ and tracking time $t_f = 2\ sec$. From these results, we can confirm that the tracking error is gradually reduced as the perturbation observers are hierarchically added.

In addition to the peak values, a performance measure is necessary to evaluate the degree of dispersion of the tracking errors. So, the $L_2$ norm of the radial tracking error is defined as $L_2[e_r] = \sqrt{(1/t_f)\int_0^{t_f} |e_r(t)|^2 dt}$. The $L_2$ norms for the above two experiments are listed in Table 2.2, where the applied gains in each case were *almost the best choice* which produced best performance without control chattering. As shown, Case 1 and Case 2 produced similar performances to one another. It means that *the FFPO not requiring feedback signals can be a good alternative to the FBPO(which is a DOB or TDC equivalent)* when a smooth reference trajectory is given. Also, Case 3 and Case 4 with two hierarchical compensation loops showed very similar results to one another but better performance than Case 1 or Case 2. However, the effect of Case 5 with a three-loop structure was not so conspicuous unlike the previous simulation result.

The results in Table 2.2 and Figures 2.10–2.13 indicate that the tracking error is greatly reduced by hierarchically adding FBPO to FFPO or SMPO to FBPO. This verifies the analysis in Theorem 2.10 and Remark 3.6.

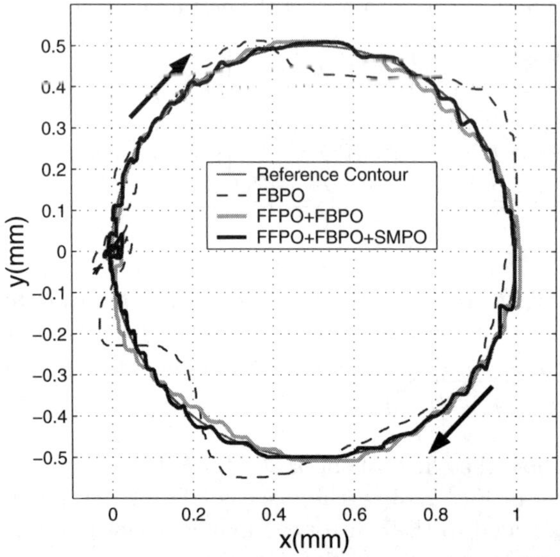


**Fig. 2.10.** (Circle($R = 500\mu m$) tracking experiment) Contour error with the sinusoidal disturbance in Fig. 2.13(b).

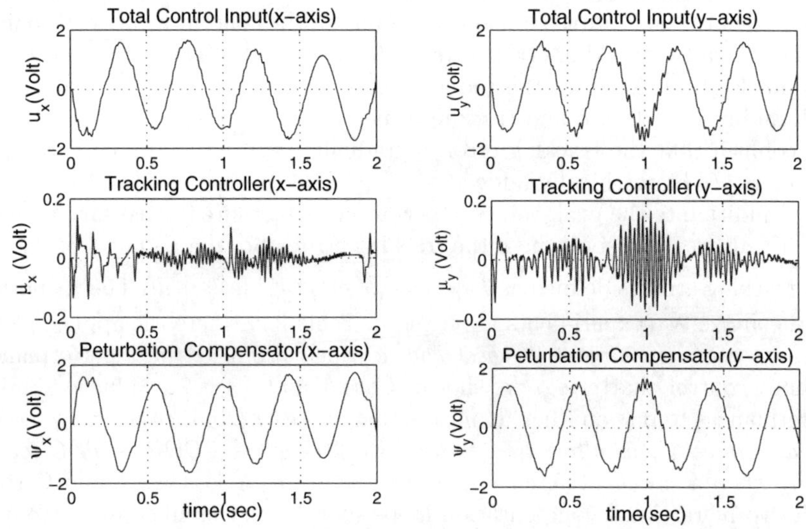


**Fig. 2.11.** (Experiment) Circle(500 $\mu m$ radius) tracking: Control inputs(Case 5), in sequence, total input $u(t)$, tracking control signal $\mu(t)$, total perturbation compensation signal, $\hat{\psi}(t)$.

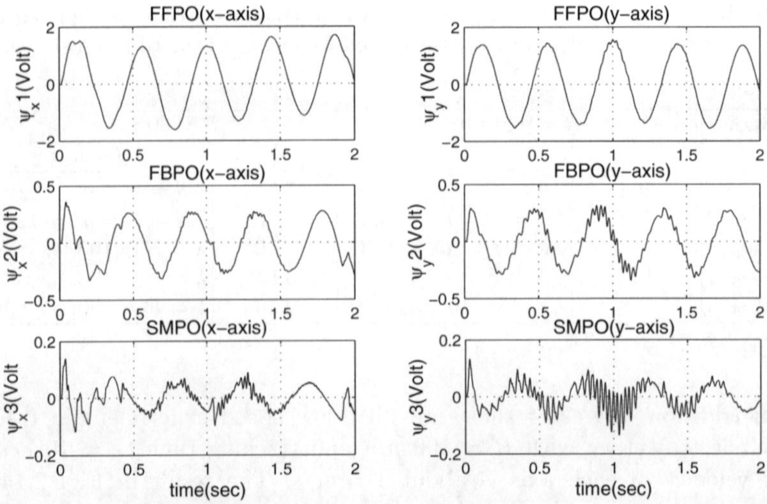


**Fig. 2.12.** (Experiment) Circle(500 $\mu m$ radius) tracking: Components of the perturbation compensation signal(Case 5), in sequence, FFPO signal, FBPO signal, and SMPO signal.

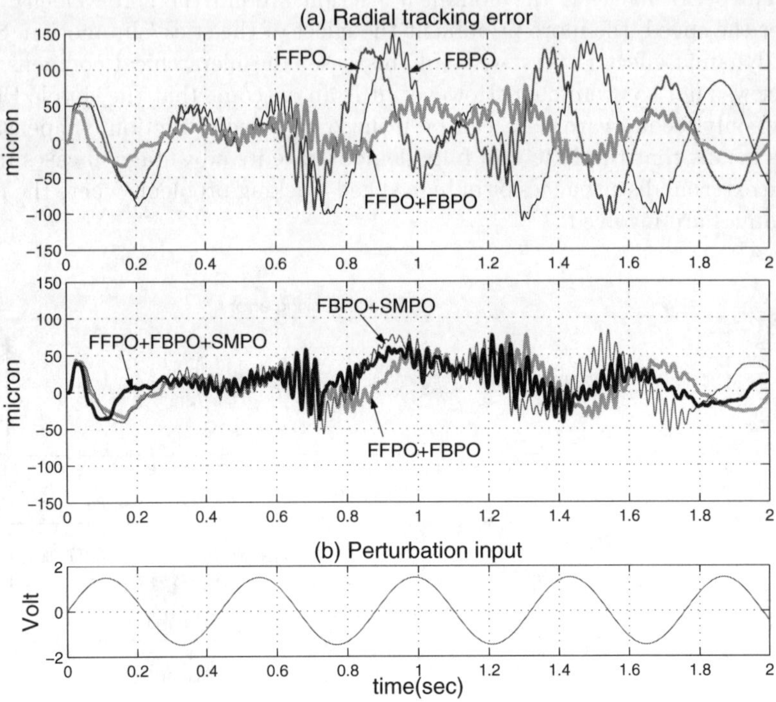


**Fig. 2.13.** (Circle($R = 30mm$) tracking experiment) (a) Radial tracking errors, (b) Sinusoidal disturbance.

**Table 2.2.** (Circle tracking experiment) Radial tracking errors and applied gains, $(\lambda, P)$: tracking controller gain, $a_c$: the cut-off frequency of the LPF, $Q(s) = a_c/(s+a_c)$.

| Radius | $R = 0.5\ mm$ | | $R = 30\ mm$ | |
|---|---|---|---|---|
| | $L_2[e_r](\mu m)$ | Gain | $L_2[e_r](\mu m)$ | Gain |
| Case 1 | 40.0 | $\lambda = P = a_c = 150$ | 53.4 | $\lambda = P = a_c = 130$ |
| Case 2 | 39.5 | $\lambda = P = a_c = 150$ | 62.6 | $\lambda = P = a_c = 120$ |
| Case 3 | 11.0 | $\lambda = P = a_c = 120$ | 27.0 | $\lambda = P = 100,\ a_c = 80$ |
| Case 4 | 14.1 | $\lambda = P = a_c = 100$ | 32.1 | $\lambda = P = 100,\ a_c = 75$ |
| Case 5 | 7.8 | $\lambda = P = 100,\ a_c = 80$ | 25.4 | $\lambda = P = 100,\ a_c = 50$ |

In addition, Fig. 2.14 shows an ultra-low speed experiment in tracking the circle trajectory with $R = 500\ \mu m$ and tracking time $t_f = 10sec$. The peak velocity of each axis is about 1 $mm/s$. The result in Fig. 2.14 was obtained by applying the FFPO(Case 1). There was no arbitrary disturbance input. As shown, the radial tracking error is smaller than about five times of one bit resolution(0.5 $\mu m$ after quadrature multiplication in decoder) of the position signal. In this ultra-low speed tracking, the major source of degrading system performance is the nonlinear friction around the zero velocity. The lower the speed, the more prominent the effect of the stick/slip motion. Since the dynamic effect is very weak in this case, the hierarchical compensation effect was not so prominent. However, it is an outcome that the simple FFPO using only feedforward data can be utilized as a useful friction compensator. This means that the HPC can fully demonstrate its advantage in cases under large external disturbances or in high speed tracking problems where the plant dynamics are involved.

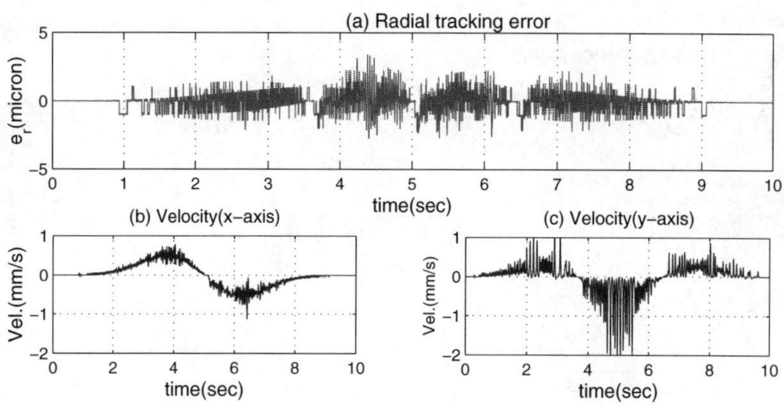


**Fig. 2.14.** (Ultra-low speed experiment) Circle($R = 500\mu m$) tracking, Applied gains: $\lambda = P = a_c = 200$.

## 2.9 Summary

In this chapter, in terms of the unified view on the class of perturbation observers, three possible points of view for perturbation compensation, i.e., FFPO, FBPO, and SMPO, were given. Moreover, by hierarchically adopting the three observers, the hierarchical perturbation compensator (HPC) which can further attenuate the residual perturbation was constructed. It was verified that the HPC enabled to get better tracking performance by using both feedforward and feedback signals through analysis and experiment. By combining the HPC and the nominal tracking controller designed using the sliding surface, the perturbation compensated sliding mode controller (PCSMC) was proposed, which can be regarded as a smooth sliding control approach in which the robust control is achieved by the perturbation observer without conventional discrete switching input. In the stability analysis, the sensitivity transfer function for the perturbation compensation error and the frequency response showed the performance enhancing capability of the HPC comparing with conventional single loop methods.

# 3

# Robust Performance of the Multiloop Perturbation Compensator

The idea of hierarchical compensation in the former chapter is extended to general $n$-loop case. In the resulting equation for stability margin of perturbation compensation loop, we meet the commonplace fact again that performance and stability should be compromised in any feedback system. The experiments on XY table and nonlinear mechanical arm demonstrate how much it is beneficial to use the perturbation compensator in respect of performance tuning, friction compensation, and getting better performance.

## 3.1 Introduction

In motion control systems for industrial applications, system performance largely depends on how intelligently the controller can compensate unpredictable perturbations such as nonlinear friction, inertia variation, and external disturbance. For example, CNC machining center is persistently affected by the cutting force and the robot manipulator in pick and place job may undergo abrupt inertia variation. In fact, the nonlinear friction around the zero velocity is the major obstacle to the fine control in most mechanical systems with contact type bearings. These perturbations make the fine positioning or tracking very difficult and also challenging. In fact, the sensitivity of the plant to such perturbations can be considerably reduced by applying a linear feedback controller. So, a simple PID controller has been successfully used for many industrial systems, but it was actually limited to the tasks in which the simple controller is just sufficient.

Above all, the *perturbation* in this chapter means the lumped uncertainty to the nominal model of a plant based on which the linear feedback controller is designed. In high speed and high precision tasks such as IC chip mounting and micro assembly, the weight of the perturbation effect to the system performance much increases and a satisfactory performance cannot be expected with the linear feedback controller alone. Hence, for robust performance of the system, an additional compensator to attenuate the time-varying perturbations and so reduce the system sensitivity to the perturbations is positively necessary. If a reliable model for a part of the perturbation is available, it can be readily compensated using a model-based technique such as friction compensator [55]. However, most of the perturbation is arbitrary and cannot be readily identified. So, we need a robust compensation method to cope with it in real-time without depending on a direct perturbation model. For this purpose, we pay attention to the *perturbation observer* based robust motion control methods in [9, 13, 31, 33, 39, 54, 55, 65, 70–72, 97, 99, 101, 105]. These controllers usually have two degrees of freedom structure composed of the conventional feedback controller for trajectory tracking and the internal perturbation observer which achieves robust performance by compensating the lumped uncertainty to the nominal plant.

A typical form of the perturbation observer is the disturbance observer (DOB) [55, 71, 97], which has been formulated for linear plants in frequency domain with output feedback condition. If we take a time-domain implementation form for the DOB except for the so-called $Q$-filter, it actually corresponds to the time delay controller (TDC) [33, 105]. The TDC has been formulated directly for nonlinear plants in time domain with the assumption that the full state is available. The internal model controller (IMC) [65] has a duality relationship with the DOB. In addition, there exist some other perturbation observers such as the adaptive robust controller(ARC) [101], the momentum feedback disturbance observer (MFDOB) [70], the model-based disturbance attenuator (MBDA) [13], and the adaptive inverse controller [99]. Commonly, these perturbation observers are based on the nominal plant model and produce compensation inputs corresponding to the deviation of the plant behavior from the nominal dynamics.

This chapter is aimed to develop an advanced perturbation compensator as an extended version of existing perturbation observers. First of all, the notion of existing perturbation observers is presented in section 2. In section 3, the concept of *residual perturbation* is introduced, in terms of which a recursive/hierarchical perturbation compensation law with multi-loop structure is developed. The performance and stability analysis in section 4 shows that the proposed multiloop perturbation compensator (MPEC) has an inherent property which can produce better performance than existing methods. The effectiveness of the proposed MPEC is confirmed through comparative experiments for the XY positioner and 2 DOF robot arm in section 5. Finally, conclusion is followed in section 6.

## 3.2 Perturbation Observer Based Robust Controller

### 3.2.1 Notion of Perturbation Observers

Although the perturbation observers such as in [13,33,55,65,71,97,99,101,105] have been formulated in some different ways, the fundamental idea is actually the same. That is, all of them include a nominal model of the plant and *indirectly* estimate the perturbation with respect to the nominal dynamics. For immediate understanding of the concept, let's consider a mass-damper system as the nominal model for a class of mechanical systems:

$$H_n\ddot{y}(t) + B_n\dot{y}(t) = u(t) + \psi(t) \tag{3.1}$$

where $(H_n, B_n)$ are nominal parameters and $u(t)$ is the control input and $\psi(t)$ is the total effective perturbation to the nominal dynamics which has the following components:

$$\psi(t) = (H_n - H(t))\ddot{y}(t) + f_c(y, \dot{y}, t) + d(t) \tag{3.2}$$

where $H(t)$ is the real inertial parameter and $d(t)$ is the external disturbance and $f_c(t)$ includes all the other plant uncertainties. From (3.1), we have an *equivalent* expression for the real perturbation (3.2) as

$$\psi_{eq} = H_n\ddot{y} + B_n\dot{y} - u. \tag{3.3}$$

Since the control interval($L$) in a digital controller is very small(usually $L < 1\ msec$), it can be reasonably assumed that $\psi(t) \approx \psi(t-L)$ if there is no extreme disturbance such as shock. Then, the perturbation at current time can be simply estimated by

$$\hat{\psi}(t) = \psi_{eq}(t-L). \tag{3.4}$$

As a result, we have a perturbation observer in a discrete form as

$$\hat{\psi}(k) = H_n\ddot{y}(k-1) + B_n\dot{y}(k-1) - u(k-1) \tag{3.5}$$

where one-step delayed signals are inevitably used to satisfy the causality between input and output sampled-data. This input will *adaptively* compensate the perturbation to the nominal dynamics. The above (3.5) can be considered as a basic implementation form of the conventional perturbation observers mentioned above without considering the filtering structure. In fact, it exactly corresponds to the time delay controller (TDC) [33,105]. For realization of (3.5), the sensor noise contained in the feedback signals should be attenuated through a low pass filter (LPF) as $\hat{\psi}_f(z) = Q(z)\hat{\psi}(z)$. When only position measure is available, the other states can be determined through a state estimator or numerical differentiation for the position output. As shown in (3.5), all the perturbation observers is based on the simple intuition that the perturbation of a plant can be effectively estimated using the nominal model(*i.e.*, internal model) and plant input/output signals.

In fact, under the identical hardware condition, the performance of the perturbation observers depends on what type of nominal model(or internal model) is adopted in the observers and what kind of low pass filter (LPF)(shown in Fig. 3.1) is used to attenuate the sensor noise. In all perturbation observers mentioned above, a form of low pass filter(*e.g.*, $Q$-filter in

DOB can be derived and it partially characterizes the difference between them. We can find some guidelines for the selection of LPF in [54,97]. Furthermore, the internal model should not be necessarily linear although a linear model is requisite for the frequency domain analysis of the perturbation observer. As we can see in time delay controllers [33,105], a nonlinear model can be readily adopted in the time domain. As described in [70], as the internal model is closer to the real plant dynamics, a better motion control performance is possible.

### 3.2.2 A Robust Motion Control Structure

The perturbation observers described above are usually adopted as an internal robust compensator in the two degrees of freedom controller. The robust motion controller of this chapter also has two DOF structure as shown in Fig. 3.1, where the tracking controller of the outer loop is designed with respect to the nominal plant and produces control inputs ($\mu$) for nominal tracking performance/stability while the perturbation compensator(or observer equivalently) generates robust control inputs ($\hat{\psi}$) to regulate the inner loop behavior to the nominal dynamics by compensating the lumped perturbation not considered in the nominal model. This two DOF structure is very efficient for robust motion controller design since the inner and outer loop can be designed independently and if necessary, the compensator of the inner loop can be readily substituted with others.

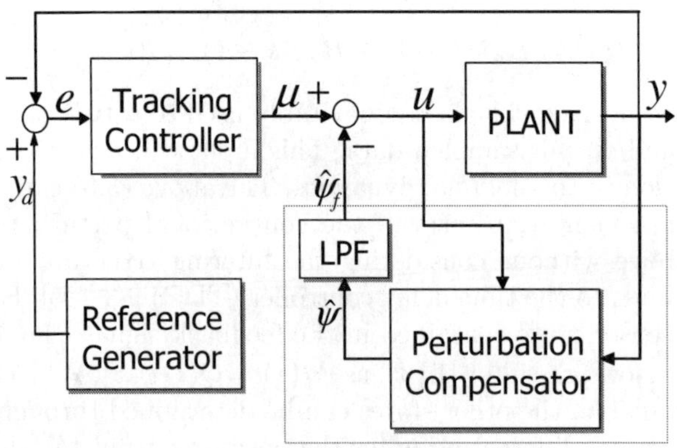


**Fig. 3.1.** Two DOF robust motion controller based on perturbation compensator.

## 3.3 A Novel Perturbation Compensation Method

### 3.3.1 Performance Limit

In the former section, it was claimed that the class of perturbation observers in [13, 33, 55, 65, 70, 71, 97, 99, 101, 105] fundamentally has the same basis and these are after all represented by (3.5) when the filtering structure is neglected. With ideal sensors, the estimation signal in (3.5) will perfectly denote the real perturbation of one-step before, *i.e.*, $\hat{\psi}(t) = \psi_{eq}(t-L) = \psi(t-L)$. However, even in this case, unavoidable is the perturbation estimation error, $\tilde{\psi} = \psi - \hat{\psi} = \psi(t) - \psi(t-L) = \Delta\psi(t)$ resulting from using one-step delayed I/O signals to construct a causal observer. Therefore, the performance of existing methods depends on the norm of the variation $\Delta\psi(t)$ and a satisfactory performance is possible on condition that the variation of the perturbation during a control interval($L$) is very small, which is a reasonable assumption in most applications. In this section, we present an efficient method to further attenuate the variation and so enhance the performance of existing perturbation observers.

While, as a trial to enhance the performance of conventional disturbance observers (DOB) [55, 71], two interesting works [13, 101] were reported which respectively has a unique design structure. First, Yao *et al.* [101] proposed the adaptive robust controller(ARC), where they put an emphasis on the point that the ARC can handle large parameter variations and it is more robust than DOB for discontinuous disturbances such as Coulomb friction. They obtained better experimental results than DOB by introducing an adaptation law. Secondly, Choi *et al.* [13] compared their model-based disturbance attenuator (MBDA) with ARC and DOB and showed better robust tracking performance for cutting process in CNC machining center. However, it was not clearly stated what property of the MBDA makes it possible to produce better performance than DOB.

### 3.3.2 Multiloop Perturbation Compensator (MPEC)

In this chapter, we propose the multiloop perturbation compensator (MPEC) in Fig. 3.2, which is composed of the main 1st loop related to the existing perturbation observers and the other auxiliary loops to compensate *residual perturbations*. In the MPEC, the perturbation is hierarchically estimated and compensated in the following manner. Above all, the 1st loop($\hat{\psi}_1$) tries to compensate the total perturbation($\psi$). The 2nd loop($\hat{\psi}_2$) does the residual perturbation due to the 1st loop(*i.e.*, the compensation error not rejected by the 1st loop) and the 3rd loop($\hat{\psi}_3$) rejects the residual perturbation due to the 1st and the 2nd loop, and so on. Then, for general $n$-loop case, the $n$th loop($\hat{\psi}_n$) will compensate the residual perturbation not attenuated in the 1st to the $(n-1)$th loop and the total compensation input is $\hat{\psi} = \sum_{i=1}^{n} \hat{\psi}_i$. Here, we define the residual perturbation for general $n$-loop case.

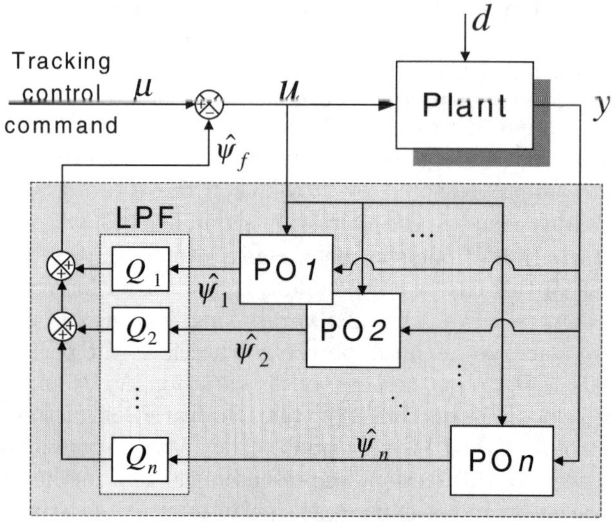


**Fig. 3.2.** Multiloop structure of the MPEC.

**Definition 3.1 (Residual Perturbation).** When the 1st($\hat{\psi}_1$) to the $(j-1)$th loop($\hat{\psi}_{j-1}$) of the MPEC in Fig. 3.2 are applied(while the other loops are cut), the perturbation compensation error is defined as the residual perturbation for the $j$th loop as follows.

$$\psi_j(t) \triangleq \tilde{\psi}_{j-1}(t) = \psi(t) - \sum_{i=1}^{j-1} \hat{\psi}_i(t) \tag{3.6}$$

where $1 \leq j \leq n$ and $\psi_1(t) = \psi(t)$. Then, the $j$th perturbation observer($PO_j$) in Fig. 3.2 will produce the estimation signal $\hat{\psi}_j(t)$ for the residual perturbation $\psi_j(t)$. □

Now, we determine the compensation law for each loop of the MPEC, $\hat{\psi}_j$ $(j = 1 \sim n)$. When the total control input, $u = \mu - \hat{\psi}$ in Fig. 3.1 is substituted into the plant (3.1), we have the relationship:

$$H_n\ddot{y}(t) + B_n\dot{y}(t) - \mu(t) = \psi(t) - \hat{\psi}(t) = \tilde{\psi}(t). \tag{3.7}$$

For convenience sake, we define the following hyperplane related to the nominal dynamics:

$$\varepsilon(t) \triangleq H_n\ddot{y}(t) + B_n\dot{y}(t) - \mu(t). \tag{3.8}$$

This function represents the deviation of the plant behavior from the nominal dynamics and keeping $\varepsilon(t) \approx 0$ means that the nominal dynamics is nearly achieved. So, the perturbation compensator in Fig. 3.1 is expected to maintain $\varepsilon(t)$ as near zero as possible.

From (3.7) and (3.8), the equivalent perturbation in (3.3) can be rewritten as $\psi_{eq}(t) = \hat{\psi}(t) + \varepsilon(t)$. Then, using the total compensation input $\hat{\psi} = \sum_{i=1}^{n} \hat{\psi}_i$, an *equivalent* expression for the residual perturbation (3.6) at $j$th loop is obtained as

$$\psi_{jeq}(t) = \psi_{eq}(t) - \sum_{i=1}^{j-1} \hat{\psi}_i(t), \tag{3.9}$$

which is composed of nominal parameters and input/output signals. Now, following the same concept in (3.4) using one step delayed signals to construct a causal observer, we have the compensation law for the $j$th loop$(1 \le j \le n)$ of the MPEC in a discrete form:

$$\hat{\psi}_j(k) = \psi_{jeq}(k-1) = \psi_{eq}(k-1) - \sum_{i=1}^{j-1} \hat{\psi}_i(k-1). \tag{3.10}$$

Then, the total sum of the perturbation compensation inputs in general $n$-loop case is determined as

$$\hat{\psi}(k) = \sum_{s=1}^{n} \hat{\psi}_s(k) = n\psi_{eq}(k-1) - \sum_{s=1}^{n-1} (n-s)\hat{\psi}_s(k-1). \tag{3.11}$$

For example, if we adopt MPEC with three loops$(n = 3)$, the compensation law of the respective loop is $\hat{\psi}_1(k) = \psi_{eq}(k-1)$, $\hat{\psi}_2(k) = \psi_{eq}(k-1) - \hat{\psi}_1(k-1)$, and $\hat{\psi}_3(k) = \psi_{eq}(k-1) - \hat{\psi}_1(k-1) - \hat{\psi}_2(k-1)$. As shown in (3.10) or (3.11), compensation inputs of one step before are recursively required to determine the current input, which characterizes the *integral property* of the MPEC.

In fact, existing perturbation observers such as disturbance observer (DOB) [55, 71] or time delay controller (TDC) [33, 105] corresponds to the single-loop case$(n = 1)$ of the MPEC. In authors' observation, the adaptive input in ARC [101] has the effect of compensating the residual perturbation and it is the basis that ARC can produce better performance than DOB in some situation. However, the performance tuning of the ARC is not so easy since it includes two gain parameters and requires the perturbation bound. In contrast, the multiloop structure of the MPEC enables to gradually reduce the perturbation compensation error according to the number of loops as far as the hardware performance allows. Also, a dual-loop compensation method can be found in [72], which is related to the two-loop case$(n = 2)$ of the MPEC.

### 3.3.3 Performance Tuning of the MPEC

From (3.6), we have $\psi_j(t) = \tilde{\psi}_{j-1}(t) = \psi_{j-1}(t) - \hat{\psi}_{j-1}(t)$ and if the equivalent perturbation $\psi_{eq}$ faithfully reconstructs the real perturbation, we can let $\hat{\psi}(t) = \psi(t-L)$ and $\hat{\psi}_{j-1}(t) = \psi_{j-1}(t-L)$ from (3.10). Then, we have

$\psi_j(t) = \psi_{j-1}(t) - \psi_{j-1}(t-L) = \Delta\psi_{j-1}(t) = \Delta^{j-1}\psi(t)$. That is, the residue of the perturbation compensation error is gradually reduced as the number of loops in the MPEC increases. For example, if we assume a 2nd order function $\psi(t) = t^2$, then $\Delta\psi = (\partial\psi/\partial t)L = 2Lt$, $\Delta^2\psi = 2L^2$, and $\Delta^3\psi = 0$. So, if we have ideal sensors, the MPEC with three loops can perfectly compensate $\psi(t) = t^2$. I.e., the system sensitivity to such perturbation is zero. On the other hand, when the conventional time delay controller (TDC) [33,105] is applied, the system will be still perturbed by the residual perturbation $\Delta\psi = 2Lt$. In addition, for a sinusoidal perturbation $\psi(t) = A\sin\omega_d t$, $\Delta\psi = LA\omega_d\cos\omega t$ and $\Delta^2\psi = -L^2A\omega_d^2\sin\omega t$. Accordingly, if the digital control frequency satisfies $f_s(= L^{-1}) > \omega_d$, the norm of the residual perturbation becomes much smaller than the original perturbation. As a result, the MPEC enables a better robust performance than existing methods, specifically, in the presence of large external disturbances.

The perturbation compensator in (3.10) or (3.11) is actually applicable only when noise-free full states are given. For practical realization of the the MPEC, the effect of sensor noise should be attenuated through low pass filters, $Q_j$-filter$(j = 1 \sim n)$ as shown in Fig. 3.2. Basically, the filters should have unity DC gains. Then, the real implementation form of (3.10) is

$$\hat{\psi}_{j_f}(k) = Q_j(z)\left(\psi_{eq}(k-1) - \sum_{i=1}^{j-1}\hat{\psi}_{i_f}(k-1)\right), \quad j = 1 \sim n \tag{3.12}$$

where $Q_j(z)$ can be considered as a linear operator corresponding to a discrete approximation of $Q_j$-filter. Then, the total compensation input is $\hat{\psi}_f = \sum_{j=1}^{n}\hat{\psi}_{j_f}$.

In fact, the perturbation suppression performance of any perturbation observer is largely dependent upon the type of the low pass filter used. For the 1st loop of the MPEC, we can adopt the popular $Q_{31}(s)$ filter with one tuning parameter:

$$Q_{31}(s) = \frac{3\tau s + 1}{(\tau s)^3 + 3(\tau s)^2 + 3\tau s + 1}, \tag{3.13}$$

which is preferred in many DOB articles [39,54,70,91] since it produces good performance. The coefficients of the $Q$-filter can be regarded as a kind of feedback gains. As the *ratio* of P, I, and D-gain is critical in the performance tuning of PID controller, the ratio of $Q$-filter coefficients is also very important to get good performance.

On the other hand, the residual perturbation in the MPEC surely becomes more noisy than the original perturbation. Hence, as the loop number increases, it would be better to use a filter with higher roll-off rate in the high frequency region. For example, we can use the filters such as

$$Q_{30}(s) = \frac{1}{(\tau s + 1)^3}, \quad Q_{40}(s) = \frac{1}{(\tau s + 1)^4}, \tag{3.14}$$

for the 2nd and the 3rd loop, which are shown in Fig. 3.3.

However, high order filters may produce adverse effect in closed loop system due to the excessive dynamic lag. So, the filters should be properly selected through the performance tuning process. When the filters are chosen, the motion control performance can be adjusted by raising the cut-off frequency($a_c = 1/\tau$) to the *allowable maximum* considering hardware limitations. In conclusion, the performance of the MPEC can be tuned as follows. *First, obtain the best performance of the 1st loop and then do fine performance tuning by adding the compensation loop one by one.*

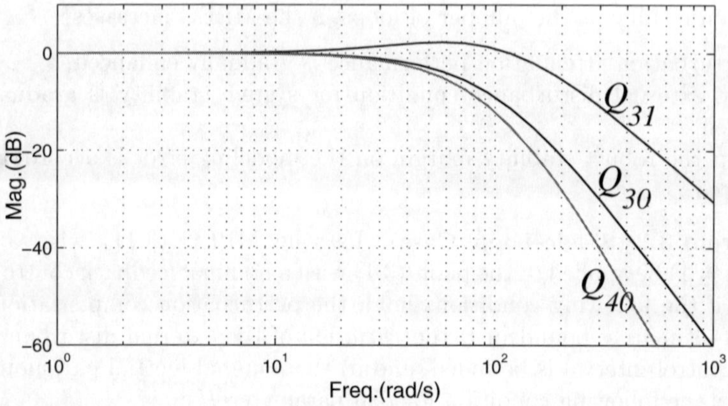


**Fig. 3.3.** Frequency response of low pass filters, $Q_{31}(s)$, $Q_{30}(s)$, and $Q_{40}(s)$ when $a_c = 1/\tau = 100$.

## 3.4 Stability/Performance Analysis

When we consider the stability analysis for the class of perturbation observers, it can be mainly classified into the frequency domain analysis for disturbance observers [9, 54, 91, 97] and the time domain analysis for time delay controllers [34, 106]. In disturbance observers, the main issue was the design of the low pass filter(i.e., $Q$-filter) which meets the internal stability based on the Youla parameterization [54, 97], passivity-based control [9], and sensitivity optimization approach [54, 91], respectively. However, the effect of time-delayed I/O signals in discrete-time implementation of disturbance observers as in (3.5) was not considered in the frequency domain analysis. While, in time delay controllers, a few stability conditions for the perturbation compensation loop were derived in time domain [34, 106]. But the effect of the low pass filter which is indispensable for the real implementation was not treated in time delay controllers with the assumption that the full state is given. In terms of remarks in the former section, the proposed MPEC can much reduce the norm of perturbation compensation error and so a greatly enhanced robust performance is possible comparing with existing methods. However, there will be certainly a limit in performance enhancement according to the increase of the compensation loops. In this section, the relationship between performance enhancement and stability robustness of the MPEC is analyzed. The analysis is conducted in discrete-time domain considering the effect of time-delayed signals in the compensation laws (3.10) and (3.11). To make the problem easy, the dynamic effect due to the $Q_j$-filter in (3.12) will be neglected in the analysis. Basically, we assume that the noise-free full state is given and the plant is time-invariant. Prior to the details, we summarize the main results: as the number of loops in the MPEC increases,

- Perturbation attenuation performance is gradually enhanced,
- The external disturbance condition for system stability is gradually relaxed,
- But, the robust stability margin on the modeling error is accordingly reduced.

**Theorem 3.2 (Single-Loop Case).** Let the MPEC (3.11) with a single loop($n = 1$) be applied to the plant (3.1) with a nominal feedback controller($\mu$). Then, if the following conditions hold, the perturbation compensation error $\tilde{\psi}(t)$ is ultimately bounded: i) the *variation* of the external disturbance during a control interval is bounded, and ii) the nominal inertial parameter($H_n$) satisfies the following condition for real parameter($H$):

$$0 < H_n < 2H. \tag{3.15}$$

*Proof.* For single-loop case($n = 1$), the MPEC (3.11) is equivalent to the basic form of conventional perturbation observers in (3.5). Above all, from (3.7), the acceleration is given as

$$\ddot{y}(k) = H_n^{-1}\left(\tilde{\psi}(k) - B_n\dot{y}(k) + \mu(k)\right). \tag{3.16}$$

From this relationship, we can expect that the time-delayed acceleration feedback term in the MPEC (3.11) or in the perturbation observer (3.5) will produce a dynamics of perturbation compensation error. If the full state is available and so the equivalent perturbation (3.3) faithfully represents the real perturbation (3.2), we can let $\psi_{eq}(k) = \psi(k)$ and $\hat{\psi}(k) = \psi_{eq}(k-1) = \psi(k-1)$. Then, using the real perturbation (3.2), the compensation error is described in a discrete form as

$$\begin{aligned}\tilde{\psi}(k) &= \psi(k) - \hat{\psi}(k) = \psi(k) - \psi(k-1) \\ &= (H_n - H)\ddot{y}(k) - (H_n - H)\ddot{y}(k-1) + \Delta f_c(k) + \Delta d(k). \end{aligned}\tag{3.17}$$

By substituting $\ddot{y}(k)$ in (3.16) into the above, the following compensation error dynamics can be derived.

$$\tilde{\psi}(k) = \left(1 - H_n H^{-1}\right) \tilde{\psi}(k-1) + \eta_1(k) \tag{3.18}$$

where the driving function $\eta_1(k)$ has the components:

$$\eta_1(k) = \left(1 - H_n H^{-1}\right) \left(B_n \Delta\dot{y}(k) - \Delta\mu(k)\right) + H_n H^{-1} \left(\Delta f_c(k) + \Delta d(k)\right) \tag{3.19}$$

where $\Delta$ denotes the variation during a control interval as $\Delta(\bullet)(k) = (\bullet)(k) - (\bullet)(k-1)$. First, we investigate the boundedness of $\eta_1(k)$ in (3.19). Since the mechanical system is a passive system, the velocity is always bounded, i.e., $\dot{y}(k) \in \mathcal{L}_\infty$. The unmodeled term $f_c(k)$ is actually composed of all bounded components such as Coriolis force, centripetal force, and nonlinear friction. So, $f_c(k) \in \mathcal{L}_\infty$. Considering a smooth change of $\dot{y}(k), f_c(k)$ in mechanical systems, their variations are also $\Delta\dot{y}(k), \Delta f_c(k) \in \mathcal{L}_\infty$ in small values. Since the tracking control input($\mu$) is designed to stabilize the nominal plant, first it can be regarded bounded, i.e., $\mu(k) \in \mathcal{L}_\infty$ and $\Delta\mu(k) \in \mathcal{L}_\infty$ under the condition that perturbation compensation loop is stable. Hence, if the *variation* of external disturbances during a control interval is bounded, i.e., $\Delta d(k) \in \mathcal{L}_\infty$, the function $\eta_1(k)$ in (3.19) will be bounded. Finally, in the discrete equation (3.18), the condition to meet $\tilde{\psi}(k) \in \mathcal{L}_\infty$ for $\eta_1(k) \in \mathcal{L}_\infty$ is readily determined as $\|1 - H_n H^{-1}\| < 1$, i.e., $0 < H_n < 2H$. So, the perturbation compensation error will be ultimately bounded as far as the condition (3.15) is satisfied. □

**Theorem 3.3 (Two-Loop Case).** Let the MPEC (3.11) with two loops($n = 2$) be applied to the plant (3.1) with a nominal tracking controller($\mu$). Then, if the following conditions hold, the perturbation compensation error $\tilde{\psi}(t)$ is ultimately bounded: i) the *double variation* of the external disturbance during a control interval is bounded, and ii) the nominal inertial parameter($H_n$) satisfies the following condition for real parameter($H$):

$$0 < H_n < \frac{4}{3}H. \tag{3.20}$$

*Proof.* In two-loop case($n = 2$), from (3.10) and (3.11), we have $\hat{\psi}(k) = \hat{\psi}_1(k) + \hat{\psi}_2(k) = 2\psi_{eq}(k-1) - \psi_{eq}(k-2)$. Based on the same reasoning as in Theorem 3.2, let $\psi_{eq}(k) = \psi(k)$. Then, the perturbation compensation error is equivalent to $\tilde{\psi}(k) = \psi(k) - \hat{\psi}(k) = \psi(k) - 2\psi(k-1) + \psi(k-2)$. According to the same procedure as in Theorem 3.2, we can derive the following error dynamics for the two-loop case.

$$\tilde{\psi}(k) = \left(1 - H_n H^{-1}\right) \left(2\tilde{\psi}(k-1) - \tilde{\psi}(k-2)\right) + \eta_2(k) \tag{3.21}$$

where the driving function $\eta_2(k)$ has the following relationship with $\eta_1(k)$ in (3.19).

$$\eta_2(k) = \eta_1(k) - \eta_1(k-1) = \Delta\eta_1(k). \tag{3.22}$$

In Theorem 3.2, we have drawn the conclusion that $\eta_1(k) \in \mathcal{L}_\infty$ for mechanical systems under the external disturbance condition, $\Delta d(k) \in \mathcal{L}_\infty$. So, from

(3.22), the function $\eta_2(k)$ will be bounded if the *double variation* of external disturbances(variation of $\Delta d(k)$) is bounded, i.e., $\Delta^2 d(k) \in \mathcal{L}_\infty$. Then, for the discrete polynomial (3.21), the sufficient condition (3.20) to satisfy $\tilde{\psi}(k) \in \mathcal{L}_\infty$ for $\eta_2(k) \in \mathcal{L}_\infty$ can be derived using the *Jury's test* [23], which is a discrete-time version of the Routh-Hurwitz criterion. □

**Theorem 3.4 (Three-Loop Case).** Let the MPEC (3.11) with three loops($n = 3$) be applied to the plant (3.1) with a nominal tracking controller($\mu$). Then, if the following conditions hold, the perturbation compensation error $\tilde{\psi}(t)$ is ultimately bounded: i) the *triple variation* of the external disturbance during a control interval is bounded, and ii) the nominal inertial parameter($H_n$) satisfies the following condition for real parameter($H$):

$$\frac{1}{2}H < H_n < \frac{8}{7}H. \tag{3.23}$$

*Proof.* In three loop case($n = 3$), from (3.10) and (3.11), we have $\hat{\psi}(k) = \hat{\psi}_1(k) + \hat{\psi}_2(k) + \hat{\psi}_3(k) = 3\psi_{eq}(k-1) - 3\psi_{eq}(k-2) + \psi_{eq}(k-3)$. Following the same procedure as in Theorem 3.2, we have the error dynamics for three-loop case:

$$\tilde{\psi}(k) = \left(1 - H_n H^{-1}(k)\right)\left(3\tilde{\psi}(k-1) - 3\tilde{\psi}(k-2) + \tilde{\psi}(k-3)\right) + \eta_3(k) \tag{3.24}$$

where the driving function $\eta_3(k)$ has the relationship:

$$\eta_3(k) = \eta_2(k) - \eta_2(k-1) = \Delta\eta_2(k) = \Delta^2\eta_1(k). \tag{3.25}$$

In Theorem 3.3, we have seen $\eta_2(k) \in \mathcal{L}_\infty$ under the condition, $\Delta^2 d(k) \in \mathcal{L}_\infty$. Then, from (3.25), the function $\eta_3(k)$ will be bounded if the *triple variation* of external disturbances(variation of $\Delta^2 d(k)$) is bounded, i.e., $\Delta^3 d(k) \in \mathcal{L}_\infty$. Similarly, by applying the *Jury's test* [23] to the discrete polynomial (3.24), the stability condition (3.23) to satisfy $\psi(k) \in \mathcal{L}_\infty$ for $\eta_3(k) \in \mathcal{L}_\infty$ can be derived. □

**Theorem 3.5 (General $n$-Loop Case).** Theorems 3.2$\sim$ 3.4 can be extended to general $n$-loop case. When the number of loops is extended to $n = 4$, we have the following error dynamics.

$$\tilde{\psi}(k) = \left(1 - H_n H^{-1}(k)\right)[4\tilde{\psi}(k-1) - 6\tilde{\psi}(k-2) + 4\tilde{\psi}(k-3) - \tilde{\psi}(k-4)] + \eta_4(k) \tag{3.26}$$

where $\eta_4(k) = \Delta\eta_3(k)$. Following the same procedure, the robust stability condition for four-loop case is given as

$$\frac{4}{5}H < H_n < \frac{14}{13}H. \tag{3.27}$$

From (3.15), (3.20), (3.23), and (3.27), we have the general relationship(but, when $n = 1$, the lower limit is 0):

$$\left[1-\frac{1}{(n-2)^2+1}\right]H < H_n < \left[1+\frac{1}{n(n-1)+1}\right]H. \tag{3.28}$$

Consequently, as $n \to \infty$, the robust stability margin on the inertia modeling error goes to zero. □

*Remark 3.6 (Relaxed External Disturbance Condition).* As we have seen in Theorems 3.2 ∼ 3.4, when the MPEC is applied, the external disturbance condition for the compensation error dynamics to be stable is gradually relaxed from $d(k) \in \mathcal{L}_\infty$ to $\Delta^n d(k) \in \mathcal{L}_\infty (n = 1, 2, 3, \cdots)$ according to the number of compensation loops. For example, if the external disturbance is a 2nd order function $d(t) = Ct^2/2$, $\Delta d(t) = (\partial d/\partial t)L = CLt$, $\Delta^2 d(t) = CL^2$, and $\Delta^3 d(t) = 0$, where $L$ is the control interval. So, the MPEC with two loops can attenuate the disturbance to constant and moreover to zero with three loops. Considering sinusoidal components $d_\omega(t) = A\exp(j\omega_d t)$ of an arbitrary disturbance, we have $\Delta^n d_\omega = -L^n A\omega_d^n \exp(j\omega_d t)$. So, $\|\Delta^n d_\omega\| < \|\Delta^{n-1} d_\omega\|$ as far as $\omega_d < L^{-1}$. □

*Remark 3.7 (Performance Enhancement).* By applying the MPEC, the perturbation to the nominal plant is changed from $\psi(t)$ in (3.2) to the residual perturbation, $\tilde{\psi} = \psi - \hat{\psi}$. From (3.7), as the norm $\|\tilde{\psi}(k)\|$ gets smaller, the behavior of the internal loop in Fig. 3.1 gets closer to the nominal plant dynamics and a better tracking performance will be produced in the outer loop of Fig. 3.1. Now, the frequency responses of the error dynamics (3.18), (3.21), and (3.24) are compared. First of all, consider the function

$$p(k) = \left(1 - H_n H^{-1}\right)\left(B_n \dot{y}(k) - \mu(k)\right) + H_n H^{-1}\left(f_c(k) + d(k)\right), \tag{3.29}$$

which is related to (3.19) as $\eta_1(k) = p(k) - p(k-1)$. Taking the $z$-transform to (3.18), (3.21), and (3.24) and considering the relationship of (3.22) and (3.25), we have the following transfer functions for the same input $p(k)$ in (3.29).

$$\text{For } n = 1, \quad \frac{\tilde{\psi}_1(z)}{p(z)} = T_1(z) = \frac{1 - z^{-1}}{1 - cz^{-1}} \tag{3.30}$$

$$\text{For } n = 2, \quad \frac{\tilde{\psi}_2(z)}{p(z)} = T_2(z) = \frac{1 - 2z^{-1} + z^{-2}}{1 - 2cz^{-1} + cz^{-2}} \tag{3.31}$$

$$\text{For } n = 3, \quad \frac{\tilde{\psi}_3(z)}{p(z)} = T_3(z) = \frac{1 - 3z^{-1} + 3z^{-2} - z^{-3}}{1 - 3cz^{-1} + 3cz^{-2} - cz^{-3}} \tag{3.32}$$

where $c = 1 - H_n H^{-1}$. The frequency responses for the above three transfer functions are shown in Fig. 3.4 when the model uncertainty is $H_n = 0.8H$ and the control interval is $L = 1$ *msec*. As shown, as the number of loops increases, the magnitude of the compensation error is rapidly reduced in the low frequency region. In Fig. 3.4, $\|T_n\| < \|T_{n-1}\|$ for the approximate frequency band $\omega < \omega_B = L^{-1}$. As a result, the norm of tracking errors in the outer loop of Fig. 3.1 will be gradually decreased according to the addition of compensation loops in the MPEC. On the other hand, the bandwidth $\omega_B$ in Fig. 3.4 is more extended as the control frequency gets faster. Then, the MPEC becomes effective for wider frequency band of input $p(k)$ and consequently, the robust performance can be more enhanced. However, the larger the inertia modeling error($\delta H = H - H_n$), the larger the high frequency peak in Fig. 3.4 in multiloop cases($n = 2, 3$). So, the trajectory of the tracking error may be more oscillatory in multiloop cases than the single-loop case although the magnitude of the amplitude is smaller. □

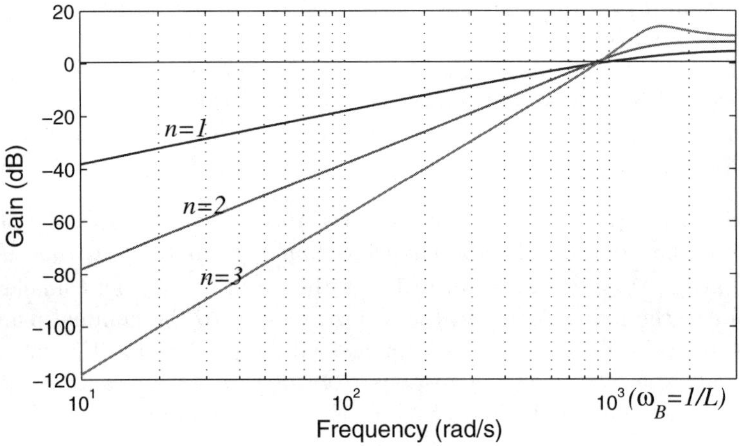


**Fig. 3.4.** Frequency response of the error transfer functions (3.30), (3.31), and (3.32) when $H_n = 0.8H$ and $L = 1$ *msec*.

*Remark 3.8 (Trade-Off and Practical Considerations).* In Remarks 3.6 and 3.7, the robust performance is accordingly enhanced as the number of loops in the MPEC increases. However, from (3.15), (3.20), (3.23), (3.27), and (3.28), it should be noted that *the robust stability margin is also gradually reduced at the cost of performance enhancement.* This result naturally illustrates the fact that performance and stability should be compromised in a control system. The robust stability condition (3.28) is just a *sufficient* condition where the effect of the $Q_j$-filters in Fig. 3.2 is not considered. If the low-pass-filtering effect is involved, the stability margin would be considerably increased(extremely, if the filter cut-off frequency is set to zero, it is equal to break the perturbation compensation loop). However, the number of loops could not be more than 3 or 4 considering the conditions (3.23) and (3.27). Furthermore, other effects in practical implementations not included in the above analysis must be considered, e.g., sensor dynamics, and limited bandwidth/resolution of digital devices. Hence, the number of loops in the MPEC should be properly limited in practical applications. □

Although a few assumptions were necessary to get above results, this kind of analysis is very helpful in understanding the property of the perturbation observer loop and the effect of the multiloop compensation. Similar stability condition to (3.15) for the single-loop case can be found in the stability analysis of time delay controllers [34, 106].

## 3.5 Application to Mechanical Systems

In this section, the proposed MPEC is applied to real plants, XY positioner and 2 DOF robot arm. Through comparative experiments, it is verified that the tracking control performance is getting better as the number of loops in the MPEC is increased and that the former analysis is valid. As a matter of fact, as well as the control algorithm, the overall performance of a control system is greatly dependent on the hardware performance such as the sampling frequency and the resolution/bandwidth of I/O devices and sensors. Therefore, in experimental results, we must pay attention to whether the proposed controller has an inherent property to improve the performance of existing controllers rather than absolute control errors and in addition, whether it provides a convenient way for performance tuning.

In the robust motion control structure shown in Fig. 3.1, the nominal tracking controller was determined based on the nominal model (3.1) as follows.

$$\mu = H_n\left(\ddot{y}_d - K_D\dot{e} - K_P e\right) + B_n\dot{y} \tag{3.33}$$

where $e = y - y_d$ is the tracking error. It is a PD controller with feedforward terms to compensate the dynamic lag between the reference($y_d$) and the output($y$) in closed loop system. As a result, a robust motion controller is determined by combining the above (3.33) with the robust perturbation compensator MPEC. Under these inputs, the closed loop system denotes the following error dynamics.

$$\ddot{e} + K_D\dot{e} + K_P e = H_n^{-1}\tilde{\psi}. \tag{3.34}$$

As shown, the overall control performance depends on the PD gains and the norm of residual perturbation $\tilde{\psi}$.

Hence, we have three tuning parameters for a motion axis, i.e., PD gains of the tracking controller (3.33) and the cut-off frequency $a_c = \tau^{-1}$ of $Q_j$-filters in the MPEC (3.12). As $Q_j$-filters, we have used $Q_1 = Q_{31}(s)$ in (3.13) for the 1st loop and $Q_2 = Q_3 = Q_{30}(s)$ in (3.14) for the 2nd and 3rd loop according to the suggestion in former section. These filters were discretely approximated through the bilinear transformation. Then, the control performance can be tuned in the following way. *First, find a gain set ($\lambda$, $a_c$) which produces a minimum tracking error in single-loop case ($n = 1$) and then apply the same values in multiloop cases ($n = 2, 3$).* As a result, the single-loop case in next experimental results corresponds to a limit of performance that can be achieved by existing perturbation observers such as DOB (but, when the same $Q_{31}$-filter is used).

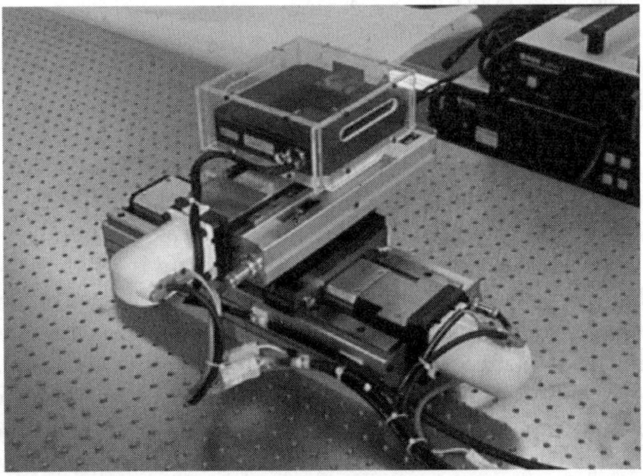

**Fig. 3.5.** Ball-screw driven XY positioner.

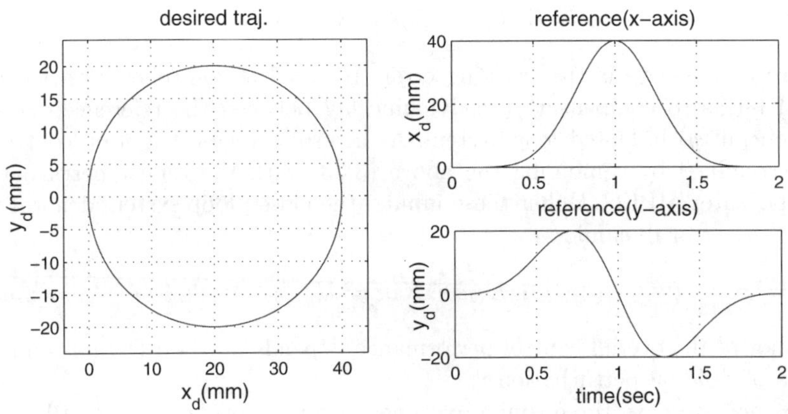


**Fig. 3.6.** A circle trajectory with radius $R = 20$ *mm* where $x(t) = R(1 - \cos\theta(t))$, $y(t) = R\sin\theta(t)$ and the rotation angle $\theta(t)$ is given as a 5th order polynomial trajectory for the target angle, $\theta_f = 2\pi$.

### 3.5.1 Control of Linear XY Table

The XY positioner in Fig. 3.5 is driven by BLDC motors where the rotary motion of the motor is converted into the linear motion through ball-screw and LM guide. It has the maximum speed 417 $mm/sec$ at 5000 RPM. The experimental setup for PC-based control is composed of Pentium processor, D/A board(12 bit), XY table, motor encoder(2500 $pulse/rev$), and counter board. The motor amplifier gain was fixed so that it generates the maximum torque($\pm 1.91$ $Nm$) for $\pm 5$ $Volt$ command. The parameters of the nominal model in (3.1) were experimentally identified under 2 $kg$ load condition as: $H_n[Volt/(m/s^2)] = (0.2020, 0.1665)$, $B_n[Volt/(m/s)] = (2.25, 1.35)$ for $(x, y)$ axis, respectively. In all experiments, the control frequency in the main processor was 1000 $Hz$ under Windows 98 OS environment. The velocity and acceleration required in the controller were determined by the backward differentiation of encoder outputs and they were passed through a 1st order low pass filter with cut-off frequency of 1000 $rad/sec$.

The control task is to robustly follow the circular trajectory in Fig. 3.6 under unpredictable large external disturbances. To clearly show the effect of the MPEC, the sinusoidal disturbance in Fig. 3.7(b) with amplitude as much as 50% of the maximum motor torque is intentionally added to the control commands of both axes. Fig. 3.8 shows the tracking errors in $(x, y)$ axis and Fig. 3.7 denotes the radial tracking error, which is defined as $e_r(t) = R - r(t)$ with $r(t) = \sqrt{x^2(t) + y^2(t)}$. In addition, Fig. 3.9 shows input components of the $x$-axis for the three-loop case($n = 3$). As shown, the tracking error was gradually decreased according as the number($n$) of loops increases. Secondly, Fig. 3.10 shows the contour error in $(x, y)$ plane in tracking the infinitesimally small circle of radius $R = 0.25$ $mm$ under the same disturbance in Fig. 3.7(b). As shown, as the number of loops increases, the trajectory gets closer to the true circle. The maximum velocities in tracking the circle of radius $R = 20$ $mm$ and $R = 0.25$ $mm$ in 2 $sec$ are about 150 $mm/sec$ and 5 $mm/sec$, respectively.

Thirdly, the chopping wave in Fig. 3.11(c) which has diverse harmonic components was inserted to both axes in tracking the circle of radius $R =$ 20 $mm$ in Fig. 3.6. In this case, the MPEC with two-loop also resulted in considerable performance improvement as shown in Fig. 3.11 but the effect of three-loop was not so noticeable. In terms of previous Remark 3.7 and Fig. 3.4, if we have a higher control frequency far beyond 1000 $Hz$, the frequency band where the MPEC with many loops is effective will be more extended. Then, even the faster harmonic components of the chopping wave can be attenuated and so the robust performance will be more enhanced by the MPEC with three loops and over.

Here, as a performance measure, we define the $L_2$ norm of tracking errors: $L_2[e_x] = \sqrt{\frac{1}{t_f}\int_0^{t_f} |e_x|^2 dt}$ for $x$-axis, $L_2[e_y] = \sqrt{\frac{1}{t_f}\int_0^{t_f} |e_y|^2 dt}$ for $y$-axis, and $L_2[e_r] = \sqrt{\frac{1}{t_f}\int_0^{t_f} |e_r|^2 dt}$ for radial direction. The $L_2$ norms resulted from above experiments are listed in Table 3.1 with applied tracking control gains

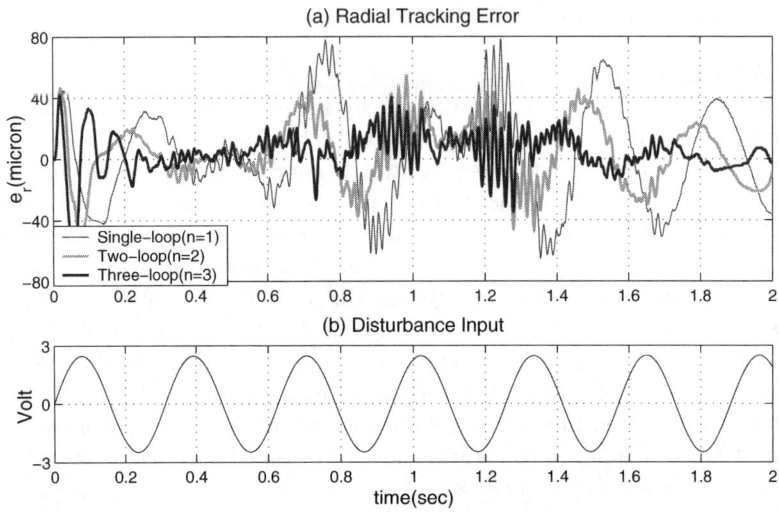


**Fig. 3.7.** (XY positioner) Circle($R = 20\,mm$) tracking experiment: (a) Radial tracking error, (b) Sinusoidal disturbance $d(t) = 2.48\sin(20t)$ $[Volt]$ of which amplitude corresponds to 50% of the maximum torque of the BLDC motor.

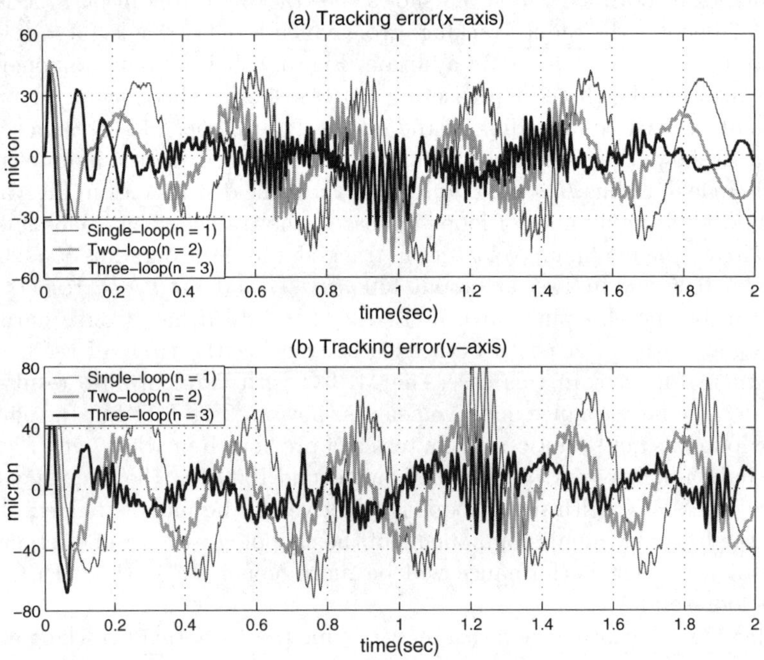


**Fig. 3.8.** (XY positioner) Circle($R = 20\ mm$) tracking experiment: (a) Tracking error in the $x$-axis, (b) Tracking error in the $y$-axis.

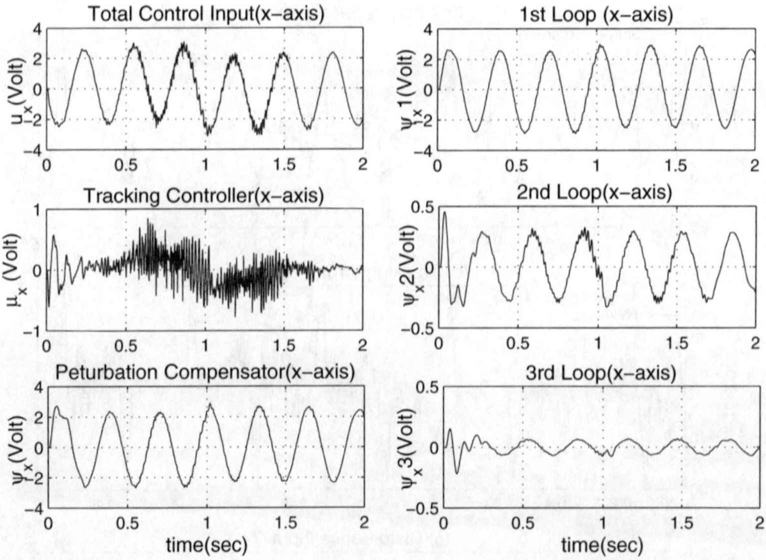


**Fig. 3.9.** (XY positioner) Circle($R = 20\ mm$) tracking experiment: Control input components of the $x$-axis in three-loop case($n = 3$).

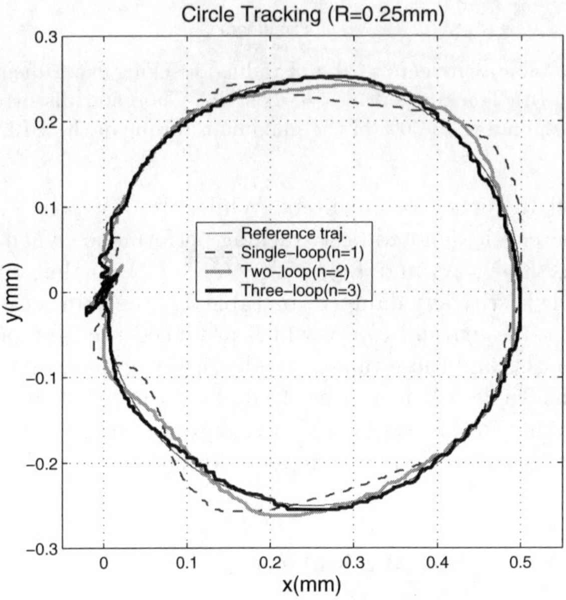


**Fig. 3.10.** (XY positioner) Circle($R = 250\ \mu m$) tracking experiment: Contour errors for the sinusoidal disturbance in Fig. 3.7(b).

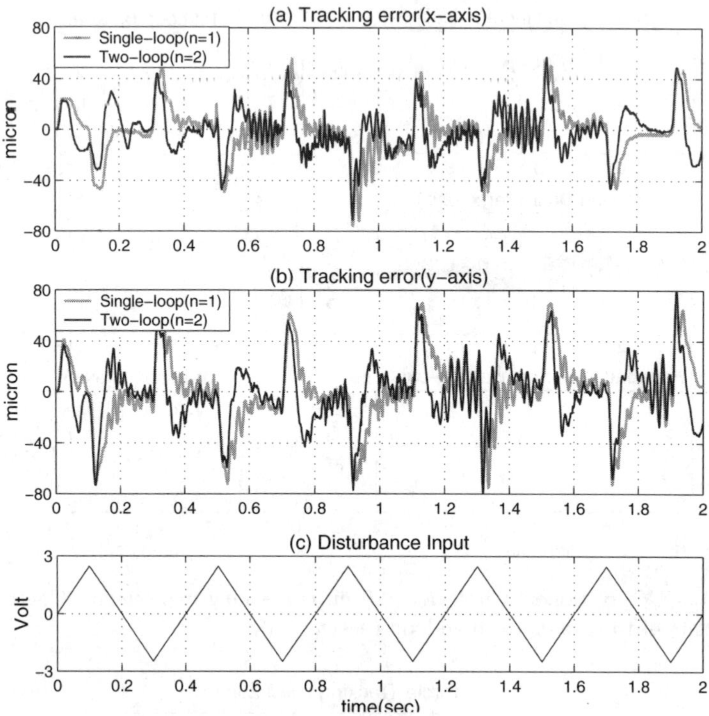


**Fig. 3.11.** (XY table) Circle($R = 20\ mm$ radius) tracking experiment: (a) Tracking error of $x$-axis, (b) Tracking error of $y$-axis, (c) Chopping disturbance with the amplitude corresponding to 50% of the maximum torque of the BLDC motor).

and the cut-off frequency of $Q_j$-filters. While, the XY positioner with high stiffness characteristic denoted best tracking performance when the PD gains are selected as $K_P = \lambda^2$ and $K_D = 2\zeta\sqrt{K_P} = 2\lambda$ so that the closed-loop system in (3.34) is critically damped. In Table 3.1, the gain set $(\lambda, a_c)$ for each axis was *almost the optimal choice* which produced the best performance in the single-loop case and these values are identically applied to other multiloop cases. From the Table 3.1, it appears that the $L_2$ norm of the tracking error is reduced roughly in an exponential way as the number of loops increases. However, it was hard to obtain any more performance enhancement with more loops over $n = 3$.

### 3.5.2 Control of Robot Manipulator

In robot manipulators, the inertial cross couplings such as Coriolis force and centripetal force become dominant in high speed motions while the nonlinear friction around zero velocity is problematic in low speed motions. These nonlinear effects are major sources which make the fine control of robot manipulators very difficult and also challenging. As a matter of fact, simple PID

**Table 3.1.** (XY positioner) Circle tracking experiment under the sinusoidal disturbance input in Fig. 3.7(b), where $R(mm)$: radius of circle, $n$: the number of loops in the MPEC, $(L_2[e_x], L_2[e_y])$: $L_2$ norm of the tracking error in $(x, y)$ axis, $L_2[e_r](\mu m)$: $L_2$ norm of radial tracking error, $(K_P = \lambda^2, K_D = 2\lambda)$: tracking control gains in (3.33), $a_c[rad/sec] = 1/\tau$: cut-off frequency of $Q_j$-filters of the MPEC.

| $R$ $(mm)$ | $n$ | $L_2[e_x]$ $(\mu m)$ | $L_2[e_y]$ $(\mu m)$ | $L_2[e_r]$ $(\mu m)$ | $x$-axis $(\lambda, a_c)$ | $y$-axis $(\lambda, a_c)$ | disturbance input |
|---|---|---|---|---|---|---|---|
| 20 | 1 | 27.5 | 38.8 | 32.1 | (185,110) | (185,100) | sine wave (Fig. 3.7(b)) |
| | 2 | 17.2 | 25.1 | 20.5 | (185,110) | (185,100) | |
| | 3 | 11.4 | 14.6 | 13.3 | (185,110) | (185,100) | |
| 0.25 | 1 | 11.6 | 16.4 | 13.4 | (295,110) | (295,100) | sine wave (Fig. 3.7(b)) |
| | 2 | 6.8 | 10.3 | 8.0 | (295,110) | (295,100) | |
| | 3 | 4.3 | 5.7 | 4.7 | (295,110) | (295,100) | |
| 20 | 1 | 20.4 | 27.7 | 23.6 | (185,110) | (185,100) | chopping wave (Fig. 3.11(c)) |
| | 2 | 17.6 | 24.3 | 20.8 | (185,110) | (185,100) | |
| | 3 | 17.7 | 24.1 | 21.0 | (185,110) | (185,100) | |

controllers have been successfully used for most industrial manipulators. However, the application is inevitably limited to simple tasks in which the PID control is sufficient. In high speed and high precision tasks such as IC chip mounting or microscopic assembly, the nonlinear effects cannot be ignored anymore and should be properly compensated by the controller.

One possible approach to attenuate the nonlinear effects is to employ the corresponding nonlinear model in the controller to produce necessary compensation torques. The conventional computed torque method [5] or the celebrated adaptive robot controller [77, 86] belong to this class. Another is to use perturbation observer techniques such as disturbance observers [31,70,71] which is very popular in recent years for robot control, where the nonlinear effect is treated as a kind of external disturbance to be rejected. The latter approach is very attractive since it requires minimal information on the plant model.

As a decentralized robot control method, the MPEC developed in this chapter was also applied to the robot arm in Fig. 3.12. It has two direct-drive revolute joints of which coordinates are $(q_1, q_2)$ and the maximum torques are $(200, 60)(N \cdot m)$, respectively. Although the direct drive arm is advantageous in motion resolution, the joint with no speed reduction gear is relatively sensitive to external forces and the structural bandwidth is much lower than the XY positioner. So, the derivative gain in the controller cannot be raised too much. The control environment was similar to the XY positioner in the former section. The nominal model used to determine the control inputs of (3.33) and (3.12) is as follows.

$$\begin{bmatrix} H_{n_1}(q_2) & 0 \\ 0 & H_{n_2} \end{bmatrix} \begin{bmatrix} \ddot{q}_1 \\ \ddot{q}_2 \end{bmatrix} = \begin{bmatrix} \tau_1 \\ \tau_2 \end{bmatrix} + \begin{bmatrix} \psi_{q1} \\ \psi_{q2} \end{bmatrix} \tag{3.35}$$

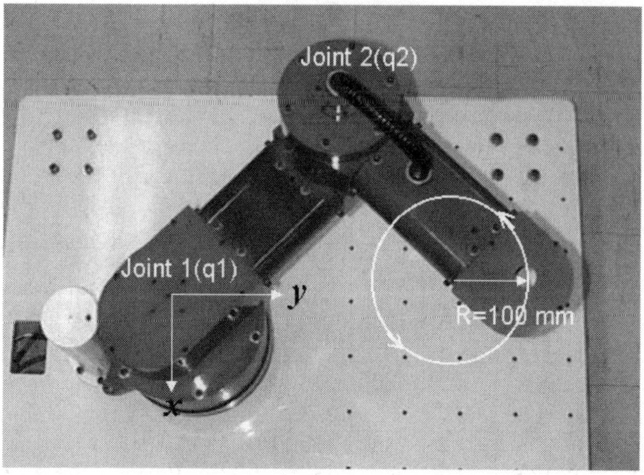


**Fig. 3.12.** Two DOF direct drive robot arm (initial position) and the desired trajectory.

That is, only the joint inertia was considered in the nominal model and all the inertial and frictional coupling effects existing in the robot dynamics were ignored. However, these neglected effects are included in the lumped perturbation $(\psi_{q1}, \psi_{q2})$ and estimated and compensated by the MPEC together with other external disturbances. As a result, we have a robust decentralized robot controller by combining the following nominal tracking controller for each joint:

$$\mu_i = H_{n_i}\left(\ddot{q}_{i_d} - K_D\dot{e}_i - K_P e_i\right), \quad (i = 1, 2) \tag{3.36}$$

with the robust compensator MPEC in (3.12) also for each joint.

Now, the robot follows the circle trajectory marked in Fig. 3.12 of which radius is $100mm$ in 2 $sec$. Along the desired trajectory, the joint inertia was varied as $(H_{n_1}, H_{n_2}) = (4.067 \sim 3.234,\ 0.483) kg \cdot m^2$. The control frequency in the experiments was commonly 1000 $Hz$. Unlike the former XY table experiments, no arbitrary disturbance was added to the control commands. Fig. 3.13 shows the change of the joint tracking error $(e_{q1}, e_{q2})$ of each joint. As shown, on the whole, the tracking error is gradually reduced according to the addition of the compensation loops in the MPEC. The case of $n = 0$ in Fig. 3.13 is the result when only the tracking input (3.36) is applied without the perturbation compensation input. The Table 3.2 lists the $L_2$ norms of the joint tracking errors of the 1st and 2nd joint and also that of radial tracking error, where the control gains and the cut-off frequency for each case were chosen as best as one can set in the performance tuning process.

From the Fig. 3.13 and the Table 3.2, it can be confirmed that the both the peak value and the $L_2$ norm of the tracking errors are obviously decreased in the two-loop case $(n = 2)$ compared with the single-loop case $(n = 1)$ but the effect of three-loop ($n = 3$) is not so conspicuous. Although the effect of multiloop compensation over $n \geq 3$ is obvious in computer simulations, the number of MPEC loops is inevitably limited in practical applications where a bandwidth exists in each hardware device. Here, we have to note that absolute values of the above tracking errors can be more reduced if the control frequency is raised to higher values and a dynamic compensation block to directly compensate the nonlinear terms in robot dynamics is added to the controller.

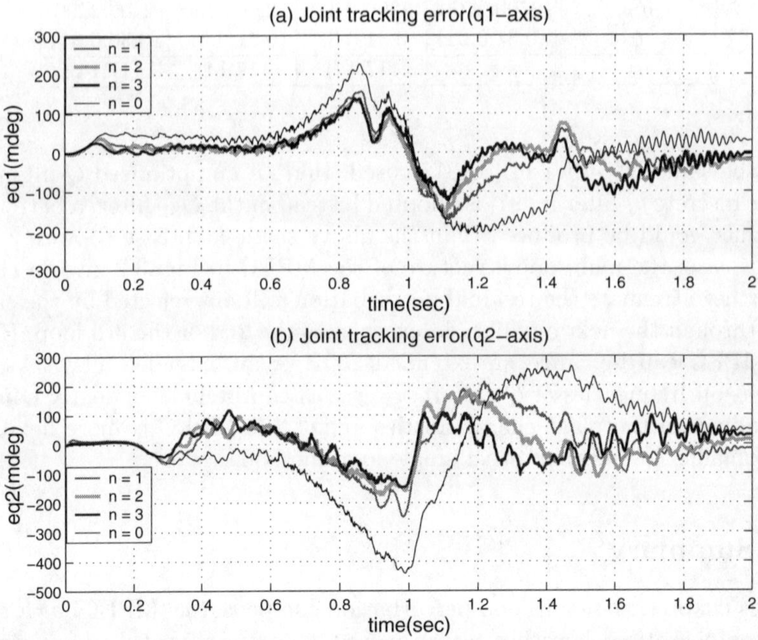


**Fig. 3.13.** (Robot arm experiment) Joint tracking errors: (a) first joint($q_1$-axis), (b) second joint($q_2$-axis).

*Remark 3.9 (Performance Comparison).* In fact, a strict performance comparison of a controller with others in numerical values is very difficult since each one has a unique gain structure. In perturbation observer based robust motion controllers, the ARC in [101] and the MBDA in [13] demonstrated a better control performance than the DOB [55, 71]. However, we can actually obtain the same performance under *fair* gain conditions. Recently, Kim et al. [41] showed that DOB, linear ARC, MBDA, and IMC [65] can be equivalently described in their robust internal-loop compensator (RIC) structure and derived equivalent $Q$-filters from them to yield the same performance. On the other hand, the single-loop case($n = 1$) in the above experimental results corresponds to the best performance of existing perturbation observers in case

**Table 3.2.** (Robot arm experiment) $n$: the number of MPEC loops, ($L_2[e_{q1}]$, $L_2[e_{q2}]$)($mdeg = deg \times 10^{-3}$): $L_2$ norms of joint tracking error for joint 1 and 2, $L_2[e_r](mm)$: $L_2$ norm of radial tracking error, ($K_P$, $K_D = 2\zeta\sqrt{K_P}$): tracking control gains in (3.36), $a_c[rad/sec] = 1/\tau$: cut-off frequency of $Q_j$-filter in the MPEC. Each pair of gain and cut-off frequency is for joint 1 and 2, respectively.

| | $n = 0$ | $n = 1$ | $n = 2$ | $n = 3$ |
|---|---|---|---|---|
| $L_2[e_{q1}](mdeg)$ | 100.2 | 57.0 | 51.2 | 50.1 |
| $L_2[e_{q2}](mdeg)$ | 183.7 | 88.2 | 73.9 | 66.9 |
| $L_2[e_r](mm)$ | 0.787 | 0.351 | 0.305 | 0.298 |
| $K_P$ | (2700,2700) | (2200,2200) | (2200,2200) | (2200,2200) |
| $\zeta$ | (0.31,0.31) | (0.31,0.31) | (0.31,0.31) | (0.30,0.31) |
| $a_c$ | × | (17,14) | (17,14) | (17,14) |

that the same $Q_{31}$-filter in (3.13) is used. Here, if an optimized $Q$-filter such as the discrete $Q$-filter in [91] is adopted instead of the $Q_{31}$-filter, a better performance could be produced than the above single-loop case. However, even in this case, the multiloop structure of the MPEC in Fig. 3.2 gives a chance to further attenuate the residual perturbation still not rejected by the single-loop through the next auxiliary loops such as the 2nd or the 3rd loop. Hence, the MPEC can be called an extended DOB or an extended TDC. As well, the cut-off frequencies of $Q_j$-filter($j = 1 \sim n$) can be independently adjusted depending on the plant characteristics. Thus, the MPEC is more flexible in performance tuning than other single-loop methods. □

## 3.6 Summary

In this chapter, the multiloop perturbation compensator (MPEC) as a recursive compensation algorithm, was proposed. It can be regarded as an $n$-degree-of-freedom controller since the total compensation input is determined by the sum of $n$ number of input components. The robust performance of the MPEC was demonstrated through the experiments. Specifically, the experiment for the robot arm showed that the MPEC could be applied as robust decentralized controller for dynamically coupled nonlinear systems. In the stability analysis for the feedback loop of the MPEC, we investigated the relationship between the robust stability and the performance enhancement according to the number of compensation loops. As a result, first, we derived a robust stability condition as a function of the inertia modeling error and the number of compensation loops, which illustrates that the stability margin with respect to the modeling error goes to zero as the number of compensation loops goes to infinity. Second, it was shown that the frequency band in which the multiloop perturbation compensation is effective is actually limited by the sampling time of the digital controller. Hence, we arrived at the conclusion that a trade-off is necessary considering the stability margin and practical hardware limitations.

# 4

# Discrete-Time Design and Analysis of Perturbation Observer

This chapter will provide some guiding principles to choose a reasonable filter for successful implementation of perturbation observer. As a design parameter of perturbation observer, the discrete $Q$-filter with so-called binomial type is considered. We put an emphasis on discrete-time analysis to examine how the stability margin and loop sensitivity are influenced depending on the type of $Q$-filter.

## 4.1 Introduction

As a robust control method for motion control systems, a class of perturbation observers such as disturbance observers [39, 44, 45, 54, 71, 97] or time-delayed controllers [34, 37, 64, 66, 106] are very popular in recent years. While linear robust control methods such as the $H_\infty$ loop shaping [17] are much mathematical and the theories cannot be readily understood by industrial engineers, the perturbation observers are physically intuitive, structurally simple, and easily implementable. The linear robust controllers make a control system very conservative due to the high-gain characteristics to cover the worst case of plant uncertainty. In contrast, the perturbation observers are characterized by the *adaptive* property to estimate the time-varying perturbation(i.e. the model uncertainty and external disturbance) and so produce just minimal control effort to compensate the perturbation.

Hence, the perturbation observer techniques could be very attractive for systems persistently perturbed by external disturbances and parameter variations. In fact, due to the advantage of simple architecture but excellent performance, perturbation observers have been widely applied to motion control systems. For example, [66] and [34] suggested time-delayed controllers (TDC) as a decentralized robot controller. [64] and [37] used the same time-delayed control techniques to compensate nonlinearities in magnetic suspension systems. Also, disturbance observers (DOB) were implemented for position/force control of robot manipulators [44,45,71] and for precision control of XY tables [39,54]. As well, adaptive robust controller (ARC) [101] and decoupled disturbance compensator (DDC) [19], which also belongs to the class of perturbation observers in authors' view, were proven to be very effective in CNC machining centers under cutting force.

To begin with, we look into the algorithm of time-delayed controller(TDC) [34,66,106] and disturbance observer(DOB) [44,54,71,97] as two representative formulations of perturbation observers, the one in time-domain and the other in frequency domain. First, let's take into account a plant model, $\dot{x} = a(x,t) + b(x,t)(u+w)$ with $\{x,a,b\} \in \mathcal{R}^{n\times 1}$ in single-input case, where the perturbation($w$) denoting the lumped uncertainty to the nominal plant satisfies the *matching condition* that the perturbation enters to the plant with control input($u$) through the same input distribution matrix $b(x,t)$. Then, by taking the pseudo-inverse of $b(x,t)$, an equivalent expression to the real perturbation is given by $w(t) = b^{+}(x,t)\left(\dot{x}(t) - a(x,t)\right) - u(t)$. The notion of the time-delayed perturbation estimation in [34,66,106] is simply to use one sampling-time($h$) delayed signals which are available at current time:

$$\hat{w}(t) = b^{+}(x,t-h)\left(\dot{x}(t-h) - a(x,t-h)\right) - u(t-h), \tag{4.1}$$

where it is basically assumed that actual perturbation changes quite smoothly between control intervals, i.e., $w(t) \approx w(t-h)$. Alternatively, if the input-output relationship of a linear plant, $y(s) = P_n(s)\left(u(s) + w(s)\right)$ is given, similarly we have $w(s) = P_n^{-1}(s)y(s) - u(s)$ through the inversion of transfer function. Then, the typical form of disturbance observer [44,54,71,97] is determined as

$$\hat{w}(s) = Q(s)\left(P_n^{-1}(s)y(s) - u(s)\right), \tag{4.2}$$

where $Q(s)$ is a low pass filter to attenuate the high frequency noise and make the mapping of $y \rightarrow \hat{w}$ causal. As shown in (4.1), TDC is a kind a static mapping of $(x, \dot{x}, u) \rightarrow \hat{w}$ where the full state is assumed to be available while DOB in (4.2) can be said as a dynamic mapping of $(y,u) \rightarrow \hat{w}$ to reconstruct the real perturbation under output feedback condition. Since the disturbance observer was formulated in terms of transfer function, the nominal plant is usually restricted to be a linear system. However, the plant model in time-delayed controllers may be linear or nonlinear due to the benefit of the time-domain approach.

In disturbance observers, the low pass filter $Q(s)$ (so-called $Q$-filter) in (4.2) was considered as a critical parameter which dominates the performance and robustness. Although some useful results on the design and analysis of the $Q$-filter were provided in [45,54,97], however, they were mainly confined to the continuous system case and it has not been clarified yet how the discretized $Q$-filter affects the performance and robustness in digital control systems differently from the continuous-time case. Specifically, the robust stability of the disturbance observer with respect to the model parameter variation has not been investigated. Moreover, the low-pass-filtering problem was not considered in the original formulations of time-delayed controllers [34, 66, 106] in spite of the importance in real implementations. On the other hand, in [39], the disturbance observer was designed in a discrete-time respect. They considered a pure inertia model with additional $m$-delay steps $z^{-m}$ to recover the phase loss of real plant and compared the sensitivity response of the discrete DOB with continuous case, through which it was shown that DOB performance could be fairly increased by reducing the mismatch between the real plant and nominal model.

In this chapter, we consider a discrete perturbation observer which actually includes time-delayed controller in (4.1) and disturbance observer in (4.2). Above all, the highlight of this chapter exists in the discrete-time analysis considering the general $m$th order discrete $Q$-filter and the inertia perturbation as a structured model uncertainty, which visualizes how the performance and robustness of the perturbation observer loop is changed according to the variation of $Q$-filter parameters, sampling time, and model parameter, respectively. Furthermore, stability conditions on time-delayed controllers [34, 106] not considering the low pass filter were proven to be specific cases of the robust stability conditions with respect to the inertia uncertainty given in this chapter.

First, a discrete perturbation observer is formulated in section 2. In section 3, we derive a perturbation compensation error dynamics in discrete-time domain, based on which robust stability conditions are determined in section 4 and sensitivity analysis is performed in section 5. Consecutively, a composite perturbation observer is proposed in section 6 to improve the perturbation attenuation performance by utilizing both feedback and feedforward signals. Section 7 shows comparative experiments. Finally, this chapter is concluded in section 8.

## 4.2 Discrete Perturbation Observer

As described in (4.1) and (4.2), two approaches for perturbation observer design are based on the same notion that perturbation of a plant can be effectively estimated using the nominal model with input/output signals. In fact, omitting the low pass filter $Q(s)$ in (4.2), the disturbance observer can be readily converted into the time-delayed controller form. Here, let's consider the issue of discrete-time implementation of two algorithms. First, the continuous transfer functions $Q(s)$ and $P_n^{-1}(s)$ in (4.2) should be discretized so that the mapping of $u \to \hat{w}$ satisfies the causality. Secondly, high frequency harmonic components in the raw signal of (4.1) beyond the actuator bandwidth should be properly filtered through a low pass filter. The discrete perturbation observer given in this section can be regarded as a discrete-time version of the disturbance observer and also an extended time-delayed controller with low pass filter.

### 4.2.1 Perturbation Model

First of all, we consider a discrete state-space plant model, $x(k+1) = A_r x(k) + B_r u(k) + B_w (f(k) + d(k))$, where $f(k)$ and $d(k)$ denote the unmodeled dynamics and external disturbance, respectively. By isolating the nominal parameters $(A, B)$ from the perturbed real parameters $A_r = A + \Delta A, B_r = B + \Delta B$, the plant model can be written as

$$x(k+1) = Ax(k) + Bu(k) + \underbrace{\Delta A x(k) + \Delta B u(k) + B_w (f(k) + d(k))}_{\text{unknown vectors}}, \quad (4.3)$$

where $x \in \mathcal{R}^{n\times 1}, \{A, \Delta A\} \in \mathcal{R}^{n\times n}, \{B, \Delta B, B_w\} \in \mathcal{R}^{n\times r}$, and $\{u, f, d\} \in \mathcal{R}^{r\times 1}$. Here, basically we assume that

- The pair $(A_r, B_r)$ and $(A, B)$ are controllable,
- The matching condition holds for the unknown (i.e., total uncertainty to the nominal model), which leads to the relationship of $\Delta A x(k) + \Delta B u(k) + B_w (f(k) + d(k)) = Bw(k)$.

Then, the above plant model can be arranged as

$$x(k+1) = Ax(k) + B(u(k) + w(k)), \quad (4.4)$$

where the perturbation vector $w \in \mathcal{R}^{r\times 1}$ has components of

$$w(k) = B^{+}(\Delta A x(k) + \Delta B u(k)) + B^{+} B_w (f(k) + d(k)). \quad (4.5)$$

Under zero-order-hold (ZOH) sampling/holding in sampled-data systems, (4.4) corresponds to the ZOH equivalent to its continuous-time counterpart and it exactly denotes continuous plant behavior at sampling instants. In designing a time-delayed controller for mechanical systems, the discrete model (4.4) has the advantage of avoiding acceleration requirement in continuous-time cases(note that $\dot{x}$ in (4.1) includes acceleration). Moreover, it makes the analysis of time-delayed controller involved sampled-data system somewhat easier since the sampling-time($h$) delay in (4.1) is easy to handle in discrete-time systems while it makes continuous-time systems infinite-dimensional.

While, by taking the pseudo-inverse of $B$ in (4.4), we can obtain an equivalent quantity to the perturbation $w(k)$:

$$w_{eq}(x, u, k) = B^{+}(x(k+1) - Ax(k)) - u(k), \quad (4.6)$$

which can be regarded as an *indirect* perturbation model to reconstruct the real perturbation in (4.5). As shown, it is composed of nominal model parameters and input/states signals.

### 4.2.2 A Discrete Perturbation Observer

In terms of the perturbation model (4.6) and by following the notion of time-delayed estimation scheme in (4.1), a causal discrete perturbation observer can be constructed as

$$\hat{w}(k) = w_{eq}(x, u, k-1). \tag{4.7}$$

However, in practical implementation point of view, the above in itself is hard to be directly applied because $B^+ (x(k+1) - Ax(k))$ part in (4.6) means the inverse plant model. Noting that usual physical plants including mechanical systems have low pass filter characteristics, the inverse model denotes a high gain characteristic at high frequencies and so the high frequency noise in (4.7) may be greatly amplified.

Hence, we must add a low pass filter to (4.7) as in the disturbance observer (4.2). However, this problem was not so definitely treated in the time-delayed controllers [34, 66, 106] although it is indispensable for practical implementation. Then, (4.7) is modified to

$$\begin{aligned} \hat{w}_f(k) &= \mathcal{Q} w_{eq}(x, u, k-1) && \text{(4.8a)}\\ &= \mathcal{Q}\left(B^+ (x(k) - Ax(k-1)) - u(k-1)\right), && \text{(4.8b)} \end{aligned}$$

where $\mathcal{Q}$ is a low-pass-filtering operator for the discrete $Q$-filter which has the following form in general $m$th order case.

$$Q_m(z) = \frac{N(z)}{D(z)} = \frac{\sum_{r=1}^{m} a_r z^{-(r-1)}}{1 + \sum_{r=1}^{m} b_r z^{-r}}, \tag{4.9}$$

where $z^{-1}$ is a delay operator and the DC gain of (4.9) should be unity, i.e., $Q_m(z = 1) = \sum_{r=1}^{m} a_r/(1 + \sum_{r=1}^{m} b_r) = 1$. In multi-input cases, we can let the matrix of low-pass-filtering operator in a diagonal form as $\mathcal{Q} = diag\{\mathcal{Q}_1, \cdots, \mathcal{Q}_r\}$.

Now, let the $Q$-filter in (4.9) be expressed as $Q_m(z) = H(zI - F)^{-1}G$, where $F$, $G$, and $H$ are constant matrices with appropriate dimensions according to the filter order. Then, the state-space realization of (4.8) can be expressed as

$$\begin{aligned} \hat{\eta}(k) &= F\hat{\eta}(k-1) + G w_{eq}(x, u, k-1) && \text{(4.10a)}\\ \hat{w}(k) &= H\hat{\eta}(k). && \text{(4.10b)} \end{aligned}$$

*Remark 4.1 (Full-state reconstruction).* As shown in (4.1) and (4.7), the time-delayed control schemes basically require full states $x(k) \in \mathcal{R}^{n\times 1}$. However, in many motion control systems, the sensory information is much restrictive and usually only position output is available. Regarding the output feedback case, the robust state estimator in [10,69] could be a good choice to reconstruct the full state, specifically when the plant uncertainty is of great concern. In many cases of mechanical systems, however, the velocity is determined as a filtered derivative of the position and the result is quite satisfactory. □

### 4.2.3 Discrete $Q$-filter

In fact, the $Q$-filter in (4.8) is not only a low pass filter to attenuate the sensor noise but also it greatly influences the stability and performance of the perturbation compensation loop. So, it has been considered as a key parameter in disturbance observer designs [45, 54, 97]. Taking some typical forms of the $Q$-filter, first, Umeno and Hori [97] employed Butterworth filters such as

$$Q(s) = \frac{1}{\tau s + 1}, \; \frac{1.41(\tau s) + 1}{(\tau s)^2 + 1.41\tau s + 1}, \; \frac{2(\tau s)^2 + 2\tau s + 1}{(\tau s)^3 + 2(\tau s)^2 + 2\tau s + 1}$$

and Ohnishi et al. [71] and Komada et al. [45] determined the $Q$-filter by the so-called Gopinath's method, which results in following forms depending on the filter order:

$$Q(s) = \frac{k_1}{s + k_1}, \; \frac{k_2 s + k_1}{s^2 + k_2 s + k_1}, \; \frac{k_3 s^2 + k_2 s + k_1}{s^3 + k_3 s^2 + k_2 s + k_1}.$$

Also, Lee [54] used binomial filters such as

$$Q_{20} = \frac{1}{(\tau s)^2 + 2\tau s + 1}, \quad Q_{21} = \frac{2\tau s + 1}{(\tau s)^2 + 2\tau s + 1},$$
$$Q_{31} = \frac{3\tau s + 1}{(\tau s)^3 + 3(\tau s)^2 + 3\tau s + 1}, \quad Q_{32} = \frac{3(\tau s)^2 + 3\tau s + 1}{(\tau s)^3 + 3(\tau s)^2 + 3\tau s + 1},$$

and generally

$$Q_{mn}(s) = \frac{1 + \sum_{r=1}^{n} c_{mr}(\tau s)^r}{1 + \sum_{r=1}^{m} c_{mr}(\tau s)^r} \quad (n < m), \tag{4.11}$$

where $1+\sum_{r=1}^{m} c_{mr}(\tau s)^r = (1+\tau s)^m$ and the coefficients have the relationship of $c_{mr} = m!/(r!(m-r)!)$. The common characteristic of the above filters is that coefficients of numerator and denominator have the same values for same order terms.

The binomial filter in (4.11) is acceptable since an easy performance tuning is possible by adjusting one parameter $\tau$ and among them $Q_{31}(s)$ filter was popularly used in many articles [39, 54, 101] since it empirically produced good performance. Therefore, let's consider (4.11) as an archetype to get real coefficients of the discrete $Q$-filter in (4.9). Now, to determine discrete counterparts for (4.11), let's approximate it by the backward difference method, i.e., $s = (1 - z^{-1})/h$, $h$: sampling time. Then, we can obtain Table 4.1, which shows coefficients of discrete $Q$-filter for 1st and 2nd order cases as functions of sampling time($h$) and filter cut-off frequency($\omega_c = 1/\tau$). In fact, the filter coefficients can be regarded as a kind of control gain set in the perturbation observer (4.8). As there exists a few gain tuning rules to determine the ratios of PID control gains, similarly, Table 4.1 can be considered as a *rule* to determine the ratios between filter coefficients.

Now, what must we choose among the filters in Table 4.1 and higher ones? At this time, we need a guideline to support the choice of a discrete filter which suitably compromises the performance and robustness. In what follows, a perturbation compensation error dynamics is derived considering the general order discrete filter in (4.9) and on this basis, robust stability conditions and sensitivity function of the perturbation observer loop are determined successively.

**Table 4.1.** Coefficients of 1st and 2nd order discrete $Q$-filter in (4.9) corresponding to the backward approximation of the binomial filter in (4.11). $\alpha = \tau/h = 1/(h\omega_c) > 1$, $h$: sampling time, and $\omega_c = 1/\tau$: cut-off frequency.

| | denominator | | numerator | | |
|---|---|---|---|---|---|
| $Q_{mn}(z)$ | $b_1$ | $b_2$ | $a_1$ | $a_2$ | $Q_{mn}(s)$ |
| $Q_{10}(z) = \frac{a_1}{1+b_1 z^{-1}}$ | $\frac{-\alpha}{1+\alpha}$ | | $\frac{1}{1+\alpha}$ | | $Q_{10}(s) = \frac{1}{\tau s+1}$ |
| $Q_{20}(z) = \frac{a_1}{1+b_1 z^{-1}+b_2 z^{-2}}$ | $\frac{-2\alpha}{(1+\alpha)}$ | $\frac{\alpha^2}{(1+\alpha)^2}$ | $\frac{1}{(1+\alpha)^2}$ | | $Q_{20}(s) = \frac{1}{(\tau s+1)^2}$ |
| $Q_{21}(z) = \frac{a_1+a_2 z^{-1}}{1+b_1 z^{-1}+b_2 z^{-2}}$ | $\frac{-2\alpha}{(1+\alpha)}$ | $\frac{\alpha^2}{(1+\alpha)^2}$ | $\frac{1+2\alpha}{(1+\alpha)^2}$ | $\frac{-2\alpha}{(1+\alpha)^2}$ | $Q_{21}(s) = \frac{2\tau s+1}{(\tau s+1)^2}$ |

Another merit of the perturbation observer is that it can be designed in a decoupled way independent of the main feedback controller and could be plugged in to existing control system to improve the performance without altering the main control architecture. In the two-degree-of-freedom motion control system in Fig. 4.1, the perturbation observer functions as a *model regulator* which rejects the lumped perturbation (4.5) to the nominal plant. With the successful working of the perturbation observer, the inner-loop will behave like a nominal model. The total control input in Fig. 4.1 is determined as the sum of main feedback control input $\mu$ and perturbation compensation input $\hat{w}$, i.e. $u = \mu - \hat{w}$. As a consequence, the control performance required in the motion control system can be robustly achieved in combination with the perturbation observer.

## 4.3 Perturbation Compensation Error Dynamics

As a matter of fact, a simple position feedback can considerably lessen the sensitivity to the perturbation. However, the most effective way to attenuate the perturbation is the acceleration feedback. The perturbation observer actually has the acceleration control property as mentioned in the disturbance observers [44,71]. In mechanical systems, two major sources of disturbing the control performance are inertia variation and nonlinear friction. On the other hand, we do not consider resonance modes due to the flexibility since the resonance mode control, if necessary, should be attacked by an additional control

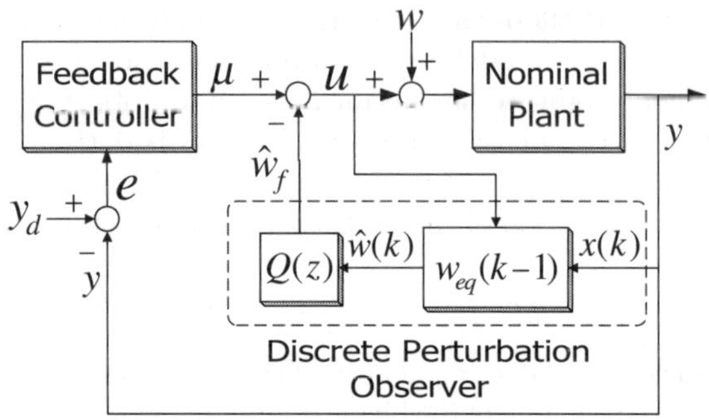


**Fig. 4.1.** A two-degree-of-freedom motion control system. $w$: lumped perturbation to the nominal plant, $y$: plant output, $x$: state variables. $w_{eq}$ is given in (4.6) and $Q(z)$ in (4.9).

law (with additional sensor/actuator in many cases). In fact, the friction of dissipative characteristics has no influence on the system stability. But, the inertia force dominantly affects the plant behavior and the inertia is actually the minimal parameter for model-based controllers.

As we know, PD control changes the damping and stiffness characteristics of the plant. In contrast, the perturbation observer will change inertial property due to the acceleration control property. Accordingly, it is very important to investigate the robust stability of the perturbation observer loop with respect to the inertia perturbation. Hence, we consider the inertia modeling error as a *structured* model uncertainty and derive a perturbation compensation error dynamics in this section when the perturbation observer (4.8) is applied to the plant (4.4). Then, in terms of the error dynamics, robustness stability conditions and sensitivity functions are determined in the next sections.

First, we consider a perturbed time-invariant pure inertia plant, $H_n\ddot{y}(t) = u(t)+w(t)$. Then, the perturbation in (4.5) is equal to $w = \Delta H\ddot{y}+f+d$, where $\Delta H(= H_n - H) \in \mathcal{R}^{r\times r}$ is the inertia modeling error($H_n$: nominal, $H$: real). Supposing that $x(k+1) = A_d x(k) + B_d\,(w(k) - f(k) - d(k))$ with $x = [y\ \dot{y}]^T$ and $y = Cx = [1\ 0]x$ is the ZOH equivalent to $\ddot{y} = \Delta H^{-1}(w - f - d)$, the perturbation can be expressed as $w(k) = B_d^+\,(x(k+1) - A_d x(k)) + f(k) + d(k)$. Now, substitute the total input $u(k) = \mu(k) - \hat{w}_f(k)$ in Fig. 4.1 into (4.4). Then, $x(k+1) - Ax(k) = B(\mu(k) + \tilde{w}(k))$, where $\tilde{w}(k) = w(k) - \hat{w}_f(k)$ is the perturbation compensation error. Noting that $A = A_d$ in pure inertia plant cases, we have the following relationship.

$$w(k) = B_d^+ B\,(\mu(k) + \tilde{w}(k)) + f(k) + d(k) \tag{4.12a}$$

$$= B_d^+ B\tilde{w}(k) + \underbrace{B_d^+ B\mu(k) + f(k) + d(k)}_{\triangleq p(k)}\,. \tag{4.12b}$$

At this point, we assume that

(A1) The outer feedback controller stabilizes the nominal plant and then, $\mu \in \mathcal{L}_\infty$.
(A2) The unmodeled dynamics and external disturbance terms in (4.5) are bounded, i.e., $\{f, d\} \in \mathcal{L}_\infty$.
(A3) The equivalent $w_{eq}(k-1)$ in (4.6) perfectly reconstructs the real perturbation one sampling-time before, i.e., $w_{eq}(k-1) = w(k-1)$.

Under (A3), the perturbation compensation error is equal to $\tilde{w}(k) = w(x,k) - \mathcal{Q}w_{eq}(k-1) = w(x,k) - \mathcal{Q}w(k-1)$. Then, by substituting (4.12) into this, we have

$$(I - B_d^+ B)\,\tilde{w}(k) = -\mathcal{Q}B_d^+ B\tilde{w}(k-1) \; + p(k) - \mathcal{Q}p(k-1). \tag{4.13}$$

Here, we restrict our attention to single-input/single-output(SISO) case to get physical insights in the analysis. Then, the matrices in (4.4) and (4.12) are given as

$$A = \begin{bmatrix} 1 & h \\ 0 & 1 \end{bmatrix}, B = \frac{1}{H_n}\begin{bmatrix} \frac{1}{2}h^2 \\ h \end{bmatrix}, B_d = \frac{1}{\Delta H}\begin{bmatrix} \frac{1}{2}h^2 \\ h \end{bmatrix} \tag{4.14}$$

where $h$ is the sampling time. From (4.14), we get $B_d^+ B = H_n^{-1}\Delta H = 1 - H_n^{-1}H$. Substituting this and discrete filter in (4.9) into (4.13), we obtain

$$\begin{aligned} &D(z)\tilde{w}(k) + N(z)H^{-1}\Delta H\tilde{w}(k-1) \\ &\qquad = H_n H^{-1}\left[D(z)p(k) - N(z)p(k-1)\right]. \end{aligned} \tag{4.15}$$

Finally, by expanding the delay operator in $D(z)$ and $N(z)$ like $z^{-r}(\bullet)(k) = (\bullet)(k-r)$, we arrive at the difference equation of perturbation compensation error:

$$\tilde{w}(k) = -\sum_{r=1}^{m} \underbrace{(b_r + a_r H^{-1}\Delta H)}_{\triangleq\, \phi_r} \tilde{w}(k-r) + \eta_m(k) \tag{4.16}$$

with

$$\eta_m(k) = H_n H^{-1}\left(p(k) + \sum_{r=1}^{m}(b_r - a_r)p(k-r)\right) \tag{4.17}$$

where $p(k)$ defined in (4.12) is bounded under (A1) and (A2), i.e., $p(k) \in \mathcal{L}_\infty$ and it has the form of

$$p(k) = H_n^{-1}\Delta H\mu(k) + f(k) + d(k). \tag{4.18}$$

Now, by expressing (4.16) in matrix form, we have

$$\begin{bmatrix} \tilde{w}(k-m+1) \\ \vdots \\ \tilde{w}(k-1) \\ \tilde{w}(k) \end{bmatrix} = \underbrace{\begin{bmatrix} 0 & 1 & \cdots & 0 \\ \vdots & \ddots & \ddots & \vdots \\ 0 & \cdots & 0 & 1 \\ -\phi_m & -\phi_{m-1} & \cdots & -\phi_1 \end{bmatrix}}_{\triangleq \Phi_m} \begin{bmatrix} \tilde{w}(k-m) \\ \vdots \\ w(k-2) \\ \tilde{w}(k-1) \end{bmatrix} + \begin{bmatrix} 0 \\ \vdots \\ 0 \\ \eta_m(k) \end{bmatrix} \tag{4.19}$$

Then, since $\eta_m(k) \in \mathcal{L}_\infty$ under $p(k) \in \mathcal{L}_\infty$, the boundedness of $\tilde{w}(k)$ is guaranteed if and only if

$$\|\Phi_m\| < 1 \Leftrightarrow |\lambda_i(\Phi_m)| < 1 \tag{4.20}$$

where $\lambda_i (i = 1, \cdots, m)$ are eigenvalues of the matrix $\Phi_m$. From $det(\lambda I - \Phi_m) = 0$, the polynomial of eigenvalue is given by

$$\lambda^m + \phi_1 \lambda^{m-1} + \cdots + \phi_{m-1}\lambda + \phi_m = 0 \tag{4.21}$$

where $\phi_r = b_r + a_r H^{-1} \Delta H,\ (r = 1, \cdots, m)$.

## 4.4 Robustness Analysis

When the real plant can be expressed with a multiplicative uncertainty as $P(s) = P_n(s)(1 + \Delta(s))$ and the complementary sensitivity function of the feedback loop is given as $T(s)$, the closed loop system will be stable if it satisfies $\|T(s)\Delta(s)\|_\infty < 1$ in terms of the result in [17]. When the disturbance observer(DOB) in (4.2) is applied, the complementary sensitivity function of the DOB loop is almost equal to the $Q$-filter, i.e., $T(s) \approx Q(s)$. Then, the robust stability condition of the disturbance observer(DOB) is given by $\|Q(s)\Delta(s)\|_\infty < 1$ [39, 54]. Although this relationship addresses a robust stability condition of the DOB loop with respect to the *unstructured* uncertainty $\Delta(s)$, this kind of frequency domain spec cannot be a practical guide for robust design and analysis since it is actually very hard to determine the explicit form of $\Delta(s)$.

In contrast, we attack the robust stability problem of the perturbation observer in discrete-time domain. From (4.16), it is evident that the stability of the perturbation compensation feedback loop is determined by the inertia modeling error $\Delta H$(a *structured* uncertainty) and the free parameter $Q_m(z)$ in (4.9). Then, as a robustness issue, to what extent is the inertia variation allowable?

First, if a 1st order discrete filter, $Q_1(z) = a_1/(1 + b_1 z^{-1})$ is used with arbitrary coefficients, the perturbation observer loop will be stable if $|-\phi_1| = \left|a_1 - b_1 - a_1 H_n H^{-1}\right| < 1$ from the stability condition (4.20). With the DC gain constraint of $Q_1(z = 1) = a_1/(1 + b_1) = 1$, this reduces to $\left|1 - a_1 H_n H^{-1}\right| < 1$. Then, the robust stability bound in regard to the inertia perturbation is given as

$$0 < H_n < \frac{2}{a_1} H. \tag{4.22}$$

Secondly, in case of 2nd order filters $Q_2(z) = (a_1 + a_2 z^{-1})/(1 + b_1 z^{-1} + b_2 z^{-2})$, (4.20) says that $|\lambda_{1,2}(\Phi_2)| < 1$ for $\lambda^2 + \phi_1 \lambda + \phi_2 = 0$. Then, after some algebraic manipulation with the DC gain constraint, $Q_2(z = 1) = (a_1 + a_2)/(1 + b_1 + b_2) = 1$, we get the following robustness bound.

$$0 < H_n < 2\left(\frac{a_1 - b_1}{a_1 - a_2}\right) H. \tag{4.23}$$

Now, substitute the coefficients of discretized binomial filters in Table 4.1 to the above (4.22) and (4.23). Then, we have Table 4.2, which shows how the stability margin of the perturbation compensation loop is changed in digital control systems depending on the sampling time($h$), filter cut-off frequency($\omega_c = 1/\tau$), and filter order($m$).

**Table 4.2.** Robust stability bounds according to the order of discretized binomial filter in Table 4.1. $\alpha = \tau/h = \omega_s/\omega_c > 1$, $h$: sampling time, $\omega_s = 1/h$: sampling frequency, and $\omega_c = 1/\tau$: filter cut-off frequency.

| filter | stability bound |
|---|---|
| $Q_{10}(z)$ | $0 < H_n < \mathcal{B}_{10} = 2(1+\alpha)H$ |
| $Q_{20}(z)$ | $0 < H_n < \mathcal{B}_{20} = 2\left(1 + 2\alpha + 2\alpha^2\right) H$ |
| $Q_{21}(z)$ | $0 < H_n < \mathcal{B}_{21} = 2\left(1 + \frac{2\alpha^2}{1+4\alpha}\right) H$ |

*Remark 4.2 (Transition of the Robustness Bound).* Comparing the upper bounds in Table 4.2, it can be seen that $\mathcal{B}_{21} \leq \mathcal{B}_{20}$ and $\mathcal{B}_{10} \leq \mathcal{B}_{20}$. This addresses that *the stability margin of the perturbation observer loop is increased as i) the sampling frequency($\omega_s = 1/h$) gets faster, ii) the cut-off frequency($\omega_c = 1/\tau$) of the filter is smaller, and iii) the filter has larger denominator order for the same numerator order and smaller numerator order for the same denominator order.* On the other hand, if the filter cut-off frequency goes to infinity, i.e., $\tau \to 0$(no filtering case), the upper bound approaches its infimum $2H$. This specific case is equal to that of the time-delay controllers [34,106] not including the low pass filter. □

## 4.5 Sensitivity Analysis

In this section, we investigate the sensitivity of perturbation observer loop with external disturbance and unmodeled dynamics and draw useful information on the discrete-time behavior of the perturbation observer in sampled-data systems by comparing the frequency responses according to the filter order($m$), sampling time($h$), and inertia modeling error($\Delta H$). First of all, from (4.16) and (4.17), we have the sensitivity function from the external disturbance($d$) or unmodeled dynamics($f$) to the perturbation compensation error($\tilde{w}$):

$$S_m(z) = \frac{\tilde{w}(z)}{d(z)} = \frac{\tilde{w}(z)}{f(z)} = \frac{H_n H^{-1}\left[1+\sum_{r=1}^{m}(b_r - a_r)z^{-r}\right]}{1+\sum_{r=1}^{m}\phi_r z^{-r}}. \tag{4.24}$$

The block diagram in Fig. 4.2 shows the inner-loop behavior when the perturbation observer (4.8) is applied to (4.4), where the perturbation compensation error is generated as a function of plant uncertainty and external disturbance through the sensitivity function $S_m(z)$. From Fig. 4.2, output transfer function of the plant under perturbation compensation is derived as

$$\frac{y(z)}{\mu(z)} = P_n(z)\left(1 + S_m(z)H_n^{-1}\Delta H\right). \tag{4.25}$$

When the outer feedback loop is closed with a feedback controller $K(z)$, the output sensitivity function is determined as

$$\frac{y(z)}{d(z)} = \frac{y(z)}{f(z)} = \frac{P_n(z)S_m(z)}{1+K(z)P_n(z)\left(1+S_m(z)H_n^{-1}\Delta H\right)}, \tag{4.26}$$

where $P_n(z) = h^2(1+z)/(2H_n(1-z)^2)$ is the ZOH equivalent of the nominal plant, $H_n\ddot{y} = u$.

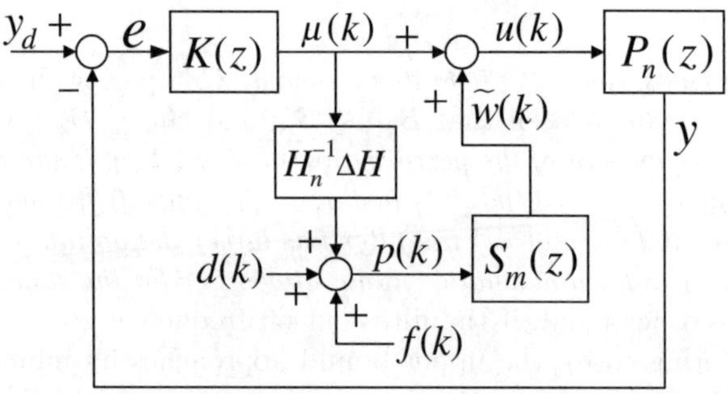


**Fig. 4.2.** Illustration of how the perturbation compensation error is generated. $f(k)$: effect of the unmodeled dynamics, $d(k)$: external disturbance input.

Above all, (4.25) confirms that if there is no modeling error(i.e., $\Delta H = 0$), the perturbation observer does not induce any phase lag in the internal feedback loop. However, supposing the model uncertainty is considerably large, high order filters would give rise to large phase fluctuation around the filter cut-off frequency. Now, we investigate the sensitivity of the system to the perturbation. First, Figures 4.3 and 4.4 are the frequency responses of (4.24). Figure 4.3 denotes the transition of the frequency response of perturbation compensation error when the family of discrete filters in Table 4.1 and the higher ones are applied and Figure 4.4 is the specific case of 3rd order filters. For higher order filters over 2nd order, their discrete forms were obtained by following the rule to make Table 4.1. Secondly, Figures 4.5 and 4.6 are the frequency responses of (4.26). Figure 4.5 shows the transition of the output sensitivity to the external disturbance or unmodeled dynamics according as the numerator order of 3rd order filters is changed and also the sampling frequency is altered in digital control systems. Figure 4.6 is the case when the inertia is perturbed as much as ±50% and the 2nd order filters are applied. While, to produce Figs. 4.5 and 4.6, a simple PD controller $K(s) = k_p + k_d s \rightarrow K(z) = (k_p + k_d/h) - (k_d/h)z^{-1}$ was used in (4.26) with $k_p = 100,\ k_d = 20$.

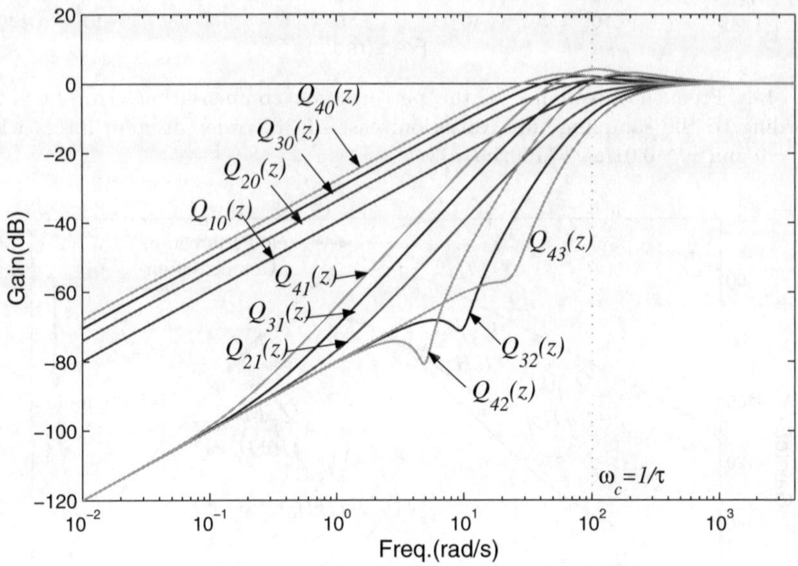


**Fig. 4.3.** Frequency response of the perturbation compensation error in (4.24) depending on the filter order when $\Delta H = 0$, $h = 0.1msec(10kHz$ sampling) and $\tau = 0.01(\omega_c = 100rad/s)$.

*Remark 4.3 (Sensitivity to Perturbation).* In Fig. 4.3, it can be seen that *the system sensitivity to the perturbation is decreased(equally, the perturbation attenuation performance is increased) as i) the denominator order of the Q-filter gets smaller for the same numerator order and ii) the numerator order gets larger for the same denominator order*, which is contrary to the result in Remark 4.2 for the robust stability bounds. It confirms that the performance

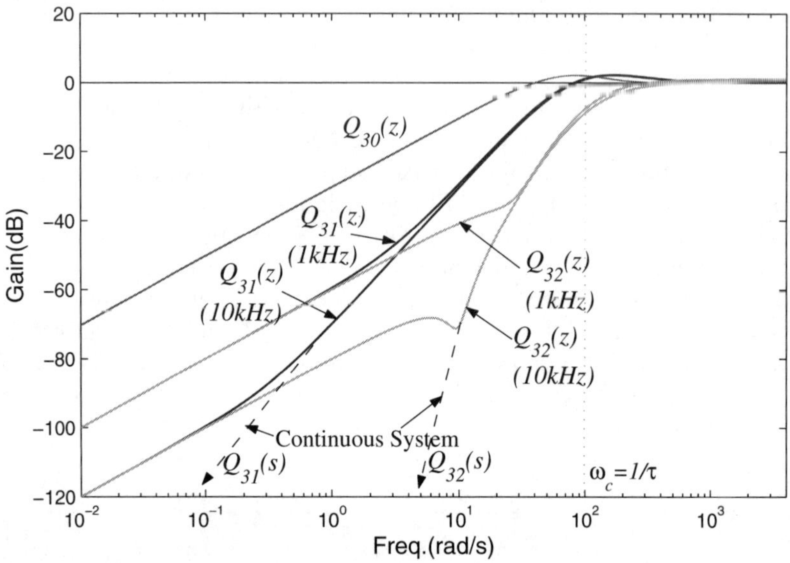


**Fig. 4.4.** Frequency response of the perturbation compensation error in (4.24) according to the sampling time variation: case of 3rd order discrete filters when $\Delta H = 0$ and $\tau = 0.01(\omega_c = 100rad/s)$.

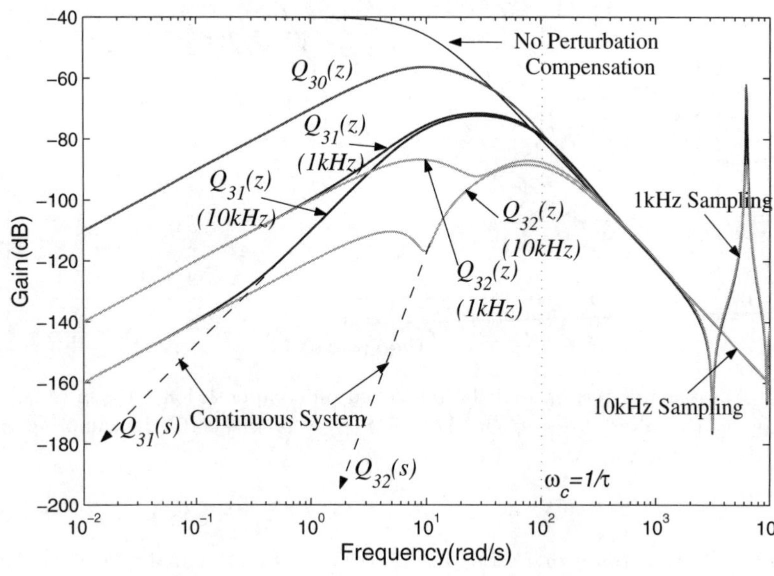


**Fig. 4.5.** Transition of the output sensitivity in (4.26) according to the sampling time variation: case of 3rd order discrete filters when $\Delta H = 0$ and $\tau = 0.01(\omega_c = 100rad/s)$.

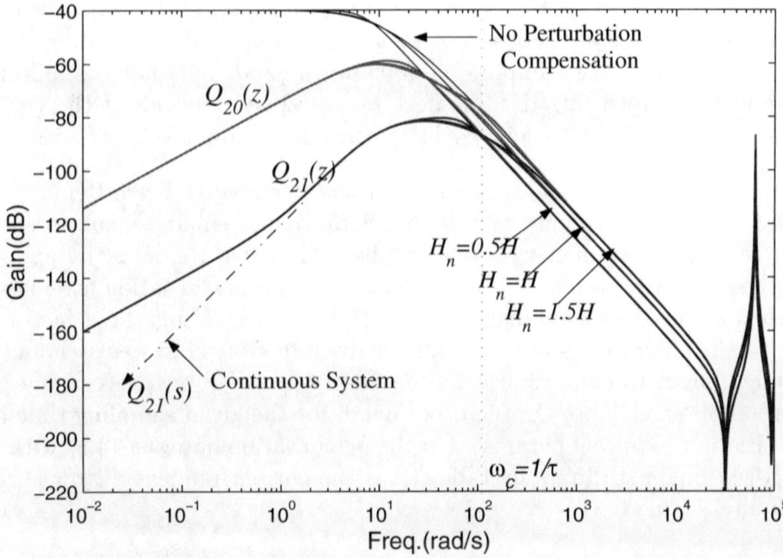


**Fig. 4.6.** Transition of the output sensitivity in (4.26) according to the inertia perturbation: case of 2nd order discrete filters, $H_n = 1$, $\Delta H = -0.5, 0, 0.5$, $h = 0.1msec$ ($10kHz$ sampling), $\tau = 0.01(\omega_c = 100rad/s)$.

in a control system has trade-offs to the stability margin. In Fig. 4.6, the output sensitivity varies with inertia perturbations only in the high frequency region beyond the filter cut-off frequency, which is physically reasonable since the performance of low frequency motions is hardly affected by the inertia variations. □

*Remark 4.4 (Effect of Time-delayed Input).* As shown in Figs. 4.3 to 4.6, the high roll-off rate below the filter cut-off frequency is maintained only in limited ranges in case of high numerator order filters and the asymptote lines are parallel to those of zero numerator order filters. This phase is resulted from the fact that the discrete perturbation observer in (4.8) uses the time-delayed input of $w_{eq}(k-1)$ with zero-order-hold(ZOH) sampling/holding and the zero-order-hold plant model is considered in deriving the difference equation in (4.13) and the transfer functions in (4.24) and (4.26). As a result, in discrete-time control systems under ZOH sampling/holding, the performance of the continuous perturbation observers [45, 54, 71, 97] will be recovered in limited bandwidths as shown in Figs. 4.3 to 4.6. □

*Remark 4.5 (Effect of Fast Sampling).* In Fig. 4.4 and 4.5, it is clear that fast sampling extends the region of high roll-off rate which overlaps with that of continuous-time systems. In other words, as the sampling frequency gets faster, the performance of the discrete perturbation observer approaches that of continuous-time case. This result is reasonable since fast sampling means small time-delay in (4.8). Hence, the sampling frequency should be sufficiently high for the perturbation observer with high numerator order filters to be effective. However, fast sampling is not always good. It is not cost effective and it may produce an adverse effect in low speed motions if the control system is equipped with a low resolution encoder without velocity sensor. □

*Remark 4.6 (Trade-off between Performance and Robustness).* From Remarks 4.2 and 4.3, the robust stability margin and performance enhancement according to the filter order conflict with one another. And from Remarks 4.4 and 4.5, even the high order filters will not produce so high perturbation attenuation performance in slow sampling systems. Hence, considering these facts, the order of the discrete $Q$-filter in (4.9) in discrete control systems should be properly chosen to compromise the performance and robustness of the perturbation observer loop. On the other hand, for the given sampling time and filter order, only tuning parameter in the perturbation observer (4.8) with the discretized binomial filters in Table 4.1 is the cut-off frequency($\omega_c = 1/\tau$) of the $Q$-filter. □

## 4.6 Composite Perturbation Observer

Within the operational bandwidth of the plant below the cut-off frequency of $Q$-filter (where $Q(z) \approx 1$), the best estimate for the current perturbation by perturbation observer in (4.7) will be the perturbation one step before, i.e., $\hat{w}(k) = w(k-1)$. However, as the perturbation $w(t)$ possesses high frequency components, its inter-sample variation, $\delta w(k) = w(k) - w(k-1)$ becomes so much larger and this will result in a large perturbation compensation error, $\tilde{w}(k) = w(k) - \hat{w}(k)$. Then if the inter-sample change can be additionally estimated, the perturbation compensation performance could be considerably enhanced. In this section, as a possible approach to fulfill this requirement, we propose a composite perturbation observer which uses both feedforward and feedback signals.

First, as an alternative to (4.8) which uses feedback signals, let's consider

$$\hat{w}_F(k) = \mathcal{Q}_F w_{eq}(x_d, u(k-1), k) \tag{4.27a}$$
$$= \mathcal{Q}_F \left(B^+ \left(x_d(k+1) - A x_d(k)\right) - u(k-1)\right) \tag{4.27b}$$

which adopts feedforward signals $x_d = [y_d \; \dot{y}_d]^T$ of a reference trajectory at sampling instants. While the former discrete perturbation observer(DPO) in (4.7) estimates the perturbation with respect to the nominal plant dynamics, the discrete *feedforward* observer(DFO) in (4.27) does with respect to the desired dynamics according to the reference trajectory. Actually, it has the advantages of no requirement of feedback signals and the preview action by the one-step ahead signal $x_d(k+1)$. Secondly, if (4.27) is applied, the plant will be still perturbed by $w_R(k) = w(k) - \hat{w}_F(k)$. This can be rewritten as

$w_R(x,u,k) = w_{eq}(x,u,k) - \hat{w}_F(k)$ using the expression of $w(k) = w_{eq}(x,u,k)$ in (4.6). Then, by applying the same time-delayed estimation technique to this, a *residue observer* is given by

$$\hat{w}_R(k) = \mathcal{Q}_R w_R(x,u,k-1) \tag{4.28a}$$
$$= \mathcal{Q}_R \left(w_{eq}(x,u,k-1) - \hat{w}_F(k-1)\right). \tag{4.28b}$$

Now, by combining (4.27) and (4.28), we have a *composite* perturbation observer(CPO) which requires both feedforward and feedback signals:

$$\hat{w}_C(k) = \hat{w}_F(k) + \hat{w}_R(k). \tag{4.29}$$

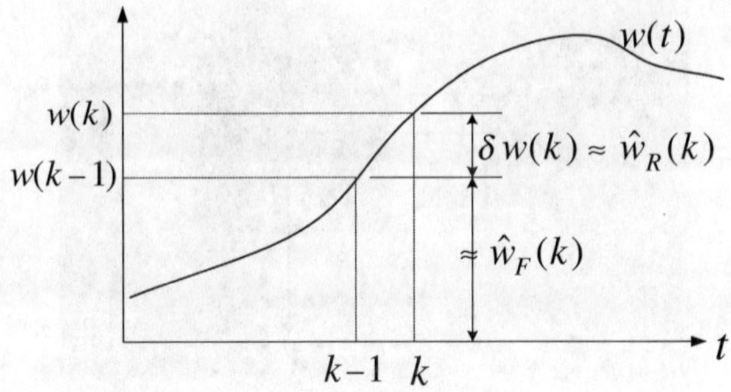


**Fig. 4.7.** Estimation of the inter-sample variation of time-varying perturbation.

Figure 4.7 illustrates how each part of (4.29) can contribute to the estimation of current perturbation $w(k)$. Roughly speaking, within the frequency band below the cut-off frequency of the low pass filters, $Q_F(z)$ in (4.27) and $Q_R(z)$ in (4.28), 1st part estimates the perturbation one step before, i.e., $\hat{w}_F(k) \approx w(k-1)$ and 2nd part approximates the inter-sample variation, i.e., $\delta\hat{w}_R(k) \approx \delta w(k)$. Hence, as the perturbation is locally closer to the linear function of time, (4.29) becomes more effective. In (4.29), simply we can let the filter as $Q_F(z) = Q_R(z)$. However, considering that raw signals before the filtering may have quite different energy spectrum, a more fine performance tuning is possible by using independent low pass filters. As an another, it should be noted that if the feedforward observer is omitted in (4.29), it is just equal to the discrete perturbation observer in (4.8) which was investigated in the previous analysis.

## 4.7 Experimental Verification

In this section, the performance of the discrete perturbation observer is experimentally verified and also the effect of the composite perturbation observer is confirmed. Experimental setup of the XY positioner in Fig. 4.8 has the closed-loop architecture for digital control: PC Pentium processor, D/A converter(12 bit), motor driver, AC servo motor(200 $W$), optical encoder(2500 $pulse/rev$), and decoder/counter(24 bit), where the control command is transferred to the motor driver every 1 $msec$ using the PC interrupt and the sampling time is also $h = 1\, msec$. The pure inertia model of each axis is respectively $H_{nx}\ddot{x} = u_x$ and $H_{ny}\ddot{y} = u_y$ where $(H_{nx}, H_{ny})[V/(m/s^2)] = (0.2020,\ 0.1665)$ and the input voltage is limited as $-5V \leq (u_x, u_y) \leq 5V$.

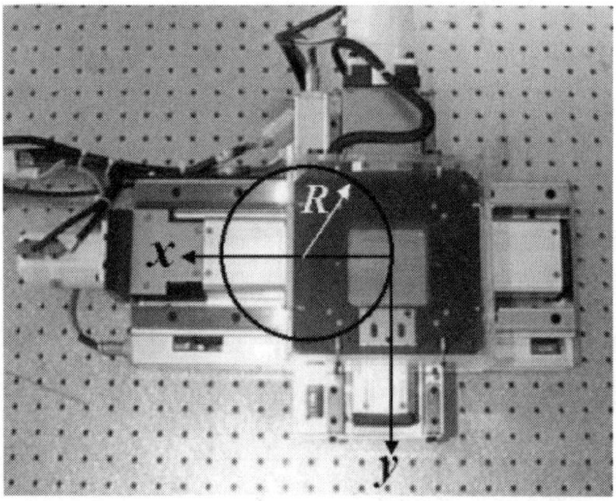


**Fig. 4.8.** XY positioner and the desired trajectory: $x(t) = R(1 - \cos\theta(t))$, $y(t) = R\sin\theta(t)$, where $R = 25mm$ and $\theta(t)$ is given by a 5th order polynomial trajectory for the terminal angle, $\theta_f = 2\pi$.

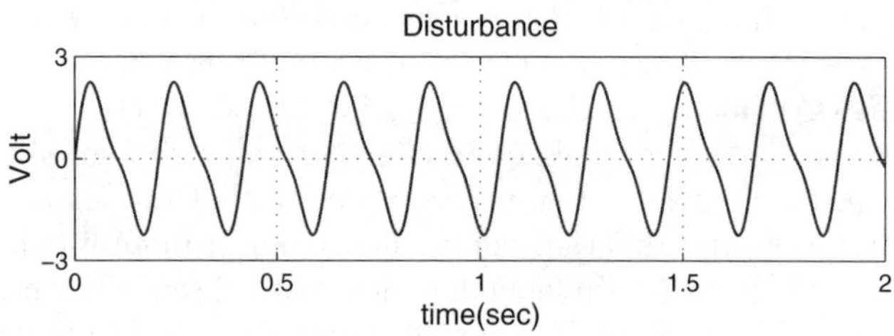


**Fig. 4.9.** External disturbance input, $d(t)[Volt] = 2sin(40t) + 0.3sin(80t)$.

The feedback controller in Fig. 4.1 is given by a simple PD rule for each axis, $\mu(k) = -K_P e(k) - K_D \dot{e}(k)$ with tracking error $e(k)$. For a pure inertia plant, we have $B^+ = (B^T B)^{-1} B^T = \beta \left[\frac{1}{2}h^2 \;\; h\right]$ with $\beta \triangleq H_n/(h^4/4 + h^2)$ in terms of (4.14) and so the equivalent perturbation in (4.6) can be written as

$$\hat{w}_{eq}(k) = \beta(0.5h^2 [x(k+1) - x(k) - h\dot{x}(k)] + h [\dot{x}(k+1) - \dot{x}(k)]) - u(k). \quad (4.30)$$

Then, the discrete perturbation observer(DPO) in (4.8) for each axis is given by

$$\begin{aligned}\hat{w}(k) =& \mathcal{Q}\hat{w}_{eq}(k-1) \\ =& \mathcal{Q}\{\beta(0.5h^2 [x(k) - x(k-1) - h\dot{x}(k-1)] \\ &+ h [\dot{x}(k) - \dot{x}(k-1)]) - u(k-1)\}.\end{aligned} \quad (4.31)$$

With 3rd order discrete filter (i.e., $m = 3$ in (4.9)), the above can be implemented as follows.

$$\hat{w}(k) = -b_1\hat{w}(k-1) - b_2\hat{w}(k-2) - b_3\hat{w}(k-3) + a_1 w_{eq}(k-1) + a_2 w_{eq}(k-2) + a_3 w_{eq}(k-3), \quad (4.32)$$

where, following the rule to make Table 4.1, the denominator coefficients of the 3rd order filters, $Q_{30}(z)$, $Q_{31}(z)$, and $Q_{32}(z)$ are commonly given as $b_1 = \frac{-3\alpha}{1+\alpha}, b_2 = \frac{3\alpha^2}{(1+\alpha)^2}, b_3 = \frac{-\alpha^3}{(1+\alpha)^3}$, but the numerator ones are respectively $a_1 = \frac{1}{(1+\alpha)^3}, a_2 = a_3 = 0$ for $Q_{30}(z)$, $a_1 = \frac{1+3\alpha}{(1+\alpha)^3}, a_2 = \frac{-3\alpha}{(1+\alpha)^3}, a_3 = 0$ for $Q_{31}(z)$, and $a_1 = \frac{1+3\alpha+3\alpha^2}{(1+\alpha)^3}, a_2 = \frac{-3\alpha-6\alpha^2}{(1+\alpha)^3}, a_3 = \frac{3\alpha^2}{(1+\alpha)^3}$ for $Q_{32}(z)$. Also, the discrete-time implementation forms of the discrete feedforward observer(DFO) in (4.27) and the composite perturbation observer (CPO) in (4.29) can be determined in similar manners. In the experiment, we used 3rd order filters and let $\mathcal{Q}_F = \mathcal{Q}_R$ in the composite perturbation observer.

The control problem is to follow a circular trajectory with radius $R = 25\ mm$ in 2 *sec* marked in Fig. 4.8. To demonstrate the robust performance, an arbitrarily large disturbance shown in Fig. 4.9 was added to the control commands of both axes. First, Figure 4.10 denotes the performance of the discrete perturbation observer(DPO) depending on the applied $Q$-filter while tracking error norms in both axes are compared in Table 4.3. As shown, the tracking performance is gradually enhanced as the numerator order is increased. It verifies Remark 4.3 and the frequency responses in Figs. 4.4 and 4.5. However, the rate of performance enhancement due to the $Q_{32}(z)$ filter is relatively small comparing with that due to the $Q_{31}(z)$ filter. As we have explained through Remark 4.5 and Figs. 4.4 and 4.5, the frequency band where the $Q$-filter of high numerator order is effective is considerably restricted by the sampling time in discrete-time systems. If a higher sampling frequency is used than $\omega_s = 1\ kHz$ as in this experiment, $Q_{32}(z)$ filter will be more effective.

Secondly, Figure 4.11 shows the performance of the composite perturbation observer (CPO) in (4.29) comparing with the discrete perturbation observer (DPO) (4.8) and the discrete feedforward observer (DFO) (4.27), where the same $Q_{31}(z)$ filter was applied. Also, tracking error norms are compared in

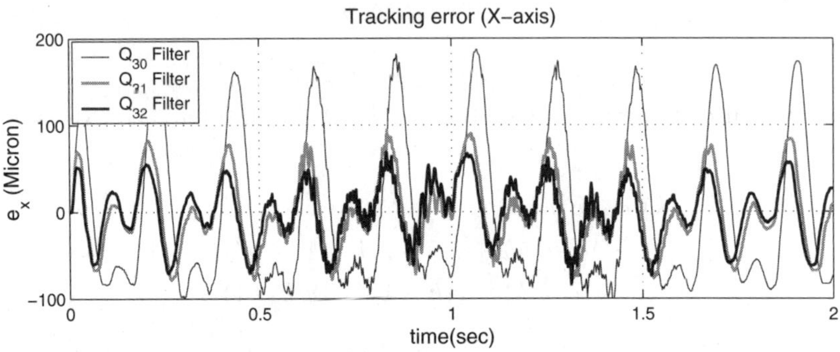


**Fig. 4.10.** Performance of the discrete perturbation observer(DPO) in (4.8) according to $Q_{30}(z)$, $Q_{31}(z)$, and $Q_{32}(z)$ filter.

Table 4.3. These results prove that the composite perturbation observer enables advanced motion controls. While, the performance in each experiment was tuned by gradually increasing the tuning parameter of the $Q$-filter (i.e. the cut-off frequency) until just before the hardware chattering happened in the XY positioner.

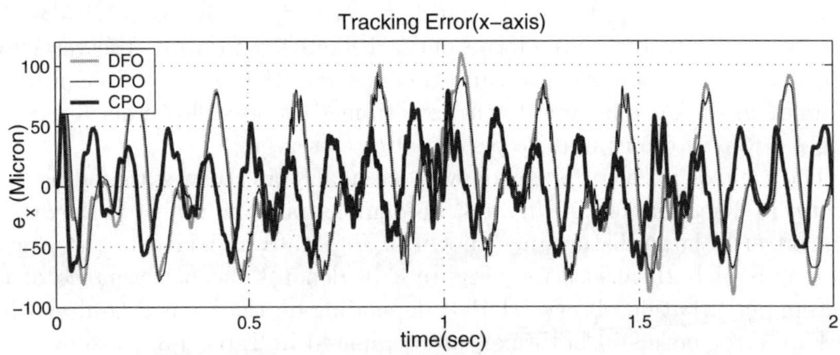


**Fig. 4.11.** Performance of the composite perturbation observer(CPO) (4.29): comparison with the discrete feedforward observer(DFO) (4.27) and the discrete perturbation observer(DPO) (4.8) when the same $Q_{31}(z)$ filter is applied.

## 4.8 Summary

In this chapter, a discrete perturbation observer was investigated. The heart of issue was to find out how the plant uncertainty and the perturbation observer parameters have an influence on the performance and stability of the perturbation compensation loop under digital control. The discrete-time analysis

**Table 4.3.** $L_2$ norm of tracking error(unit: micron).

| | Observer | $L_2[e_x]$,$L_2[e_y]$ | filter | $(\omega_{cx}, \omega_{cy})$ |
|---|---|---|---|---|
| | | 89.5, 89.8 | $Q_{30}(z)$ | (220,215) |
| Fig. 4.10 | DPO (4.8) | 43.3, 44.2 | $Q_{31}(z)$ | (130,130) |
| | | 33.8, 38.4 | $Q_{32}(z)$ | (65,60) |
| | DFO (4.27) | 46.6, 48.7 | | (170,165) |
| Fig. 4.11 | DPO (4.8) | 43.3, 44.2 | $Q_{31}(z)$ | (130,130) |
| | CPO (4.29) | 31.0, 34.6 | | (115,110) |

– $L_2[e_x] = \sqrt{(1/t_f)\int_0^{t_f} e_x^2 dt}$, $L_2[e_y] = \sqrt{(1/t_f)\int_0^{t_f} e_y^2 dt}$
– $e_x = x - x_d$, $e_y = y - y_d$,
– $(\omega_{cx}, \omega_{cy})$: $Q$-filter cut-off frequency for $(x, y)$ axis
– While, PD control gains of $(x, y)$ axis are fixed as
$K_P = \lambda^2$, $K_D = 2\zeta\lambda$ $(\lambda = 70, \zeta = 0.45)$

considering general $m$th order discrete $Q$-filter clarified how the stability robustness and sensitivity of the perturbation observer involved discrete control systems were varied according to the change of filter parameters (numerator order, denominator order, and cut-off frequency), sampling frequency, and model parameter (inertia) perturbation. The difference equation of perturbation compensation error derived in this chapter was the clue to lead to the robustness and sensitivity analysis, where the robust stability conditions and the discrete frequency responses suggested useful guidelines to support the design of discrete perturbation observer with moderate discrete $Q$-filter.

# 5

# Combined Observer Design for Robust State Estimation and Kalman filtering

The former chapters dealt with how we can achieve robust motion control with the help of perturbation observer. Now, as an antithesis topic, we discuss robust state estimation problem on the basis of perturbation observer, explaining what we are able to get in performance respect but what will be lost in respect of noise amplification. In the combined observer framework of this chapter, the long time criticism on the white noise assumption in Kalman filter can be quite a bit mitigated and the modified Kalman filter could be accepted as a practical solution to have reliable estimates when the process noise is non-white.

## 5.1 Introduction

State estimation problem has been one of the critical issues in control theory and applications. A reliable state estimate for a process is indispensable not only for control purpose but also for other applications such as navigation of spacecraft, monitoring, and fault diagnosis in mechanical systems. As is well-known, given the measurements $y = g(x, u)$ for the process $\dot{x} = f(x, u)$, it is possible to reconstruct the full-state by the estimator $\dot{\hat{x}} = f(\hat{x}, u) + \mathcal{K}(y - \hat{y})$ with innovation process [25]. However, since the mathematical model $(f, g)$ is only an approximation to the physical process and the actual plant is usually affected by external disturbances, there always exists a discrepancy between the real states and estimates. Like the other feedback systems, the uncertainty effect not considered in the plant model could be significantly reduced by increasing the estimator gain $\mathcal{K}$. However, those with innovation process alone certainly have a limit to cover all uncertainties with wide ranges in a frequency spectrum. As a simple example, conventional state estimators such as Luenberger observer or Kalman filter produce biased estimates when the plant is under biased external disturbances.

In fact, the accuracy of the state estimation largely depends on how well the physics of a real plant is copied to the estimator structure. One thing that should be remembered is that when the actuator of a real plant is saturated, the same saturation input must be applied to the state estimator. From a control point of view, the perturbation (i,e., model uncertainty and external disturbance) to the nominal plant should be rejected but in a state estimation respect, it should be possibly added to the estimator to increase the estimation accuracy by reducing the gap between actual plant behavior and nominal dynamics in the estimation algorithm. Hence, for robustness of state estimation performance, it is desirable to adopt an additional function to reflect the unknown perturbation in the observer structure.

A number of robust state observers have the form of standard observer plus a robustifying function. For example, Slotine et al. [84] and Walcott and Zak [98] designed sliding mode observers with additional switching function to address plant uncertainties in terms of the Lyapunov stability. Moura et al. [69] and Chang et al. [10] suggested a function to estimate plant uncertainties and added it to the state estimator. Gu and Poon [27] determined a nonlinear function to guarantee a Lyapunov stability of the state observer for a linear plant with parameter perturbation. For stochastic systems, on the other hand, Petersen and McFarlane [75] and Shaked and Souza [80] investigated robust Kalman filters which guarantee the boundedness of state estimation error for norm-bounded uncertainties. As a different approach, Tsui [96] and Hou et al. [32] tried to find a transformation to decouple the state observer from the effect of unknown inputs. Most of the above works provide mathematical rigor for a given problem statement. However, some of them have drawbacks in respect of implementation practicality. For example, an upper bound of uncertainty is needed to construct the state observers or linear filters in [27, 75, 80] and it would be a challenging work to find such decoupling transformations in [32,96] although they exist. In contrast, the robust state observers in [10, 69, 84, 98] have the merit of not requiring information on the plant uncertainty itself.

Following the idea to enhance the state estimation accuracy by supplementing an robustifying function to the conventional observer, a *combined observer* is proposed in this chapter, which is comprised of linear state estimator and the discrete perturbation observer(PO) in [51]. The combined observer is not only a robust state estimator but also an output feedback perturbation observer, where the full-state reconstruction and perturbation observation are performed simultaneously. Related to the combined observer setting, we have two fundamental issues: i) How much can the perturbation observer contribute to the state estimation accuracy according to the design parameter variation when it is involved in the state estimator? ii) Reciprocally, how does the perturbation observer performance change when the state estimator is used instead of full-state feedback? These problems will be investigated through the sensitivity analysis and numerical examples. Besides, the idea of combined observer synthesis can be extended to stochastic systems. Related to this, Kim and Oh [42] showed an example that Kalman filtering performance could be considerably increased by introducing a disturbance detector in the conventional Kalman filter. In section 4, discrete Kalman filter is reformulated with perturbation observer. As a result, we have a modified recursive algorithm which additionally includes the perturbation estimate update process.

In a sense, this chapter is highly inspired by the works in [10, 69, 84, 98]. However, the discrete-time design and analysis considering the coupled dynamics between the state estimator and perturbation observer is completely new. First of all, in section 2, a recursive estimation algorithm of the combined observer is formulated and the coupled estimation error dynamics and sensitivity functions are derived. In section 3, the combined observer is applied to 2nd order mechanical systems, where a design procedure, sensitivity analysis, numerical examples, and experimental results are presented. Section 4 is devoted to robust Kalman filtering for stochastic systems. Finally, this chapter is concluded in section 5.

## 5.2 Combined State Estimator–Perturbation Observer

### 5.2.1 A Combined Observer Synthesis

Consider a discrete state-space plant model,

$$x_{k+1} = Ax_k + Bu_k + \Gamma w_k, \tag{5.1a}$$

$$y_k = Cx_k + v_k \tag{5.1b}$$

with $x_k \in \Re^n$ state, $u_k \in \Re^r$ control input, $y_k \in \Re^l$ measurement, $w_k \in \Re^p$ perturbation to the linear nominal plant (i.e., unknowns composed of nonlinearities, uncertainties, and external disturbances), $v_k \in \Re^l$ measurement noise, and $A$, $B$, $\Gamma$, and $C$ known constant matrices with appropriate dimensions. Here, we assume that the perturbation is bounded ($w \in \mathcal{L}_\infty$) and admissible (i.e., the magnitude and the frequency spectrum are within the control actuator capability).

For the plant (5.1), a *combined observer* is defined as follows:

$$\hat{x}_{k+1} = A\hat{x}_k + Bu_k + L(y_k - C\hat{x}_k) + \Gamma\hat{w}_k, \tag{5.2}$$

where the first three terms denote the standard Luenberger observer (with innovation process for the residual $\tilde{y}_k = y_k - \hat{y}_k$) and the perturbation observer of the last term is to estimate the unknown ($w_k$) in (5.1). From (5.1) and (5.2), the estimation error dynamics is given by

$$\tilde{x}_{k+1} = (A - LC)\tilde{x}_k - Lv_k + \Gamma\tilde{w}_k, \tag{5.3}$$

where $\tilde{x}_k = x_k - \hat{x}_k$ and $\tilde{w}_k = w_k - \hat{w}_k$. That is, the perturbation term in the error dynamics is changed from $\Gamma w_k$ to $\Gamma\tilde{w}_k$ comparing with Luenberger observer case. Hence, so much enhancement of the estimation accuracy can be expected accordingly as the norm $\|\tilde{w}_k\|$ decreases.

In [51], we have constructed a discrete perturbation observer as an extension of disturbance observers [71,97] and time-delayed controllers [33,66,105], which has the following form:

$$\hat{w}_k = \mathcal{Q} \cdot w_{eq}(x, k-1) \;\; \text{with} \tag{5.4a}$$

$$w_{eq}(x, k-1) = \Gamma^{+} \left( x_k - A x_{k-1} - B u_{k-1} \right), \tag{5.4b}$$

where $w_{eq}(x,k)$ is a perturbation equivalent which can be determined by the plant model (5.1), $\Gamma \in \Re^{n \times p}$ has a left inverse $\Gamma^{+} = (\Gamma^T \Gamma)^{-1} \Gamma^T$ when $n = \dim(x_k) > p = \dim(w_k)$, and $\mathcal{Q} = \mathrm{diag}\{Q_1, \cdots, Q_p\}$ is a low-pass-filtering linear operator to cut off high frequency harmonics in (5.4b). The low pass filters could have the following general $m$th order form:

$$Q_i(z) = \frac{\sum_{k=1}^{m} b_k z^{-(k-1)}}{1 - \sum_{k=1}^{m} a_k z^{-k}}, \quad i = 1, \cdots, p \tag{5.5}$$

but with unity DC gain, $Q_i(1) = 1$.

*Remark 5.1.* When the perturbation observer is applied as a robust controller in addition to the main feedback controller, the perturbation compensation input to the control actuator can be determined as $u_k = -B^{+} \Gamma \hat{w}_k$. This is possible only when the perturbation in (5.1) satisfies the matching condition, i.e., $\mathrm{rank}[B, \Gamma] = \mathrm{rank}[B]$. But, this constraint is no longer required when the perturbation observer is involved in the state estimator for just estimation purpose. □

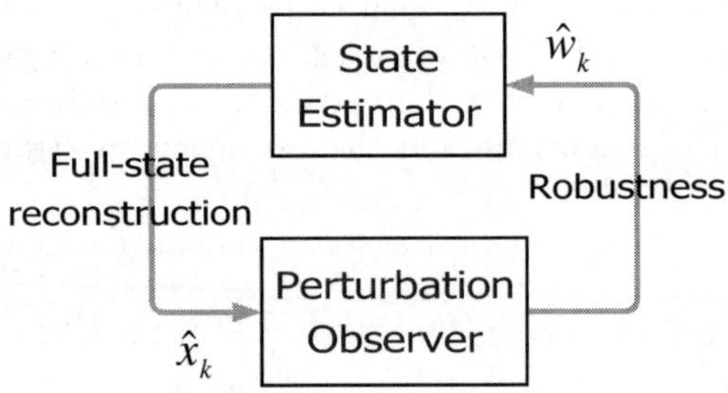


**Fig. 5.1.** Dynamical inter-connection between state estimator and perturbation observer in the combined observer.

Let the state-space realization form of (5.4) with $\mathcal{Q}(z) = H(zI - F)^{-1}G$ be

$$\hat{\eta}_k = F \hat{\eta}_{k-1} + G w_{eq}(x, k-1), \;\; \hat{w}_k = H \hat{\eta}_k, \tag{5.6}$$

where $F$, $G$, and $H$ are constant matrices with appropriate dimensions according to the filter order. Note that $\eta \in \Re^q$ and $q = \sum_{i=1}^{p} m_i$ with $m_i$ the order of

$Q_i$ filter$(i = 1, \cdots, p)$ and if a 1st order filter is used in (5.4), then $q = p$ and $H = I_p$. As shown, the full-state is needed to implement (5.4). Now, the perturbation observer under output-feedback condition where a state estimator is required can be written as

$$w_{eq}(\hat{x}, k-1) = \Gamma^+ \left(\hat{x}_k - A\hat{x}_{k-1} - Bu_{k-1}\right), \tag{5.7a}$$

$$\hat{\eta}_k = F\hat{\eta}_{k-1} + Gw_{eq}(\hat{x}, k-1), \tag{5.7b}$$

$$\hat{w}_k = H\hat{\eta}_k. \tag{5.7c}$$

Then, (5.7a)–(5.7c) and (5.2) form a recursive estimation formula of the combined state estimator–perturbation observer. As a consequence, in the combined observer setting, state estimator and perturbation observer are reciprocal with one another. That is, as the schematic shown in Fig. 5.1, the state estimator requires a perturbation observer for robustness of state estimation accuracy and conversely, the perturbation observer needs a state estimator for full-state reconstruction.

### 5.2.2 Stability

From (5.1b), (5.2), and (5.7), we get the relationship:

$$\begin{aligned} w_{eq}(\hat{x}, k) &= \Gamma^+ \left(\hat{x}_{k+1} - A\hat{x}_k - Bu_k\right) = \Gamma^+ \left(LC\tilde{x}_k + Lv_k + \Gamma H\hat{\eta}_k\right) \\ \text{and} \quad \hat{\eta}_{k+1} &= (F + GH)\hat{\eta}_k + G\Gamma^+ L \left(C\tilde{x}_k + v_k\right). \end{aligned}$$

Since $\hat{\eta}_k = \eta_k - \tilde{\eta}_k$ from the definition of $\tilde{w}_k = w_k - \hat{w}_k$, the following is given:

$$\tilde{\eta}_{k+1} = -G\Gamma^+ LC\tilde{x}_k + (F + GH)\tilde{\eta}_k + \eta_{k+1} - (F + GH)\hat{\eta}_k - G\Gamma^+ Lv_k.$$

Then, together with (5.3), we have a coupled estimation error dynamics:

$$\begin{bmatrix} \tilde{x}_{k+1} \\ \tilde{\eta}_{k+1} \end{bmatrix} = \underbrace{\begin{bmatrix} A - LC & \Gamma H \\ -G\Gamma^+ LC & F + GH \end{bmatrix}}_{\triangleq A_o} \begin{bmatrix} \tilde{x}_k \\ \tilde{\eta}_k \end{bmatrix} + \begin{bmatrix} 0 \\ \eta_{k+1} - (F + GH)\eta_k \end{bmatrix} - \begin{bmatrix} L \\ G\Gamma^+ L \end{bmatrix} v_k \tag{5.8}$$

As shown, the state estimator and the perturbation observer are dynamically coupled in the combined observer set and when $x \in \Re^n$ and $\eta \in \Re^q$, the stability condition is given by

$$|\lambda_i(A_o)| < 1, \ (i = 1, \cdots, n + q) \tag{5.9}$$

with $\lambda_i(A_o)$ eigenvalues of $A_o$ in (5.8). Hence, *the low pass filter in the perturbation observer (5.4) not only should be stable, it also should satisfy the overall stability condition (5.9).* Then, for the given plant $(A, \Gamma, C)$, the estimator gain $L$ in (5.2) and the low pass filter parameters $(F, G, H)$ in (5.7) can be determined to satisfy (5.9) through the eigenvalue assignment design procedure.

At this point, we consider the situation that a feedback controller is closed with the combined observer (5.2). By substituting $u = -K\hat{x}$ into (5.1) and using $\hat{x}_k = x_k - \tilde{x}_k$, we get $x_{k+1} = (A - BK)x_k + BK\tilde{x}_k + \Gamma w_k$. By combining this with (5.8), the closed loop dynamics can be written as

$$\begin{bmatrix} x_{k+1} \\ \tilde{x}_{k+1} \\ \tilde{\eta}_{k+1} \end{bmatrix} = \underbrace{\begin{bmatrix} A - BK & BK & 0 \\ 0 & A - LC & \Gamma H \\ 0 & -G\Gamma^+ LC & F + GH \end{bmatrix}}_{\triangleq A_{cl}} \begin{bmatrix} x_k \\ \tilde{x}_k \\ \tilde{\eta}_k \end{bmatrix}$$
$$+ \begin{bmatrix} \Gamma H \eta_k \\ 0 \\ \eta_{k+1} - (F + GH)\eta_k \end{bmatrix} - \begin{bmatrix} 0 \\ L \\ G\Gamma^+ L \end{bmatrix} v_k, \quad (5.10)$$

from which immediately we have $\det(\lambda I - A_{cl}) = \det(\lambda I - A + BK) \cdot \det(\lambda I - A_o)$. Hence, the *separation principle* [6] for the observer-based controller still holds and the combined observer poles can be assigned without considering feedback controller.

### 5.2.3 Sensitivity Functions

Taking $z$-transforming to the error dynamics (5.8), we get

$$\begin{bmatrix} \tilde{x}(z) \\ \tilde{\eta}(z) \end{bmatrix} = (zI - A_o)^{-1} \left\{ \begin{bmatrix} 0 \\ zI - F - GH \end{bmatrix} \eta(z) - \begin{bmatrix} L \\ G\Gamma^+ L \end{bmatrix} v(z) \right\}, \quad (5.11)$$

where the inverse of the partitioned matrix $(zI - A_o)$ can be readily derived as follows by using matrix inversion lemmas[1]:

$$(zI - A_o)^{-1} = \begin{bmatrix} \hat{\Delta}^{-1} & \hat{\Delta}^{-1}\Gamma H(zI - F - GH)^{-1} \\ -(zI - F - GH)^{-1} G\Gamma^+ LC\hat{\Delta}^{-1} & \Delta^{-1} \end{bmatrix} \quad (5.12)$$

where

$$\hat{\Delta} = (zI - A + LC) + \Gamma H(zI - F - GH)^{-1} G\Gamma^+ LC \quad (5.13)$$
$$\Delta = (zI - F - GH) + G\Gamma^+ LC(zI - F - GH)^{-1}\Gamma H. \quad (5.14)$$

Then, (5.11) is arranged as

$$\begin{bmatrix} \tilde{x}(z) \\ \tilde{\eta}(z) \end{bmatrix} = \begin{bmatrix} S_{\tilde{x}\eta}(z) & S_{\tilde{x}v}(z) \\ S_{\tilde{\eta}\eta}(z) & S_{\tilde{\eta}v}(z) \end{bmatrix} \begin{bmatrix} \eta(z) \\ v(z) \end{bmatrix} \quad (5.15)$$

with four kinds of sensitivity functions:

[1]See Appendix

$$\mathcal{S}_{\tilde{x}\eta}(z) = \hat{\Delta}^{-1}\Gamma H, \tag{5.16}$$
$$\mathcal{S}_{\tilde{x}v}(z) = -\hat{\Delta}^{-1}\left[I + \Gamma H(zI - F - GH)^{-1}G\Gamma^{+}\right]L, \tag{5.17}$$
$$\mathcal{S}_{\tilde{\eta}\eta}(z) = \Delta^{-1}(zI - F - GH), \tag{5.18}$$
$$\mathcal{S}_{\tilde{\eta}v}(z) = \left[(zI - F - GH)^{-1}G\Gamma^{+}LC\hat{\Delta}^{-1} - \Delta^{-1}G\Gamma^{+}\right]L, \tag{5.19}$$

which denote, in sequence, the sensitivity of the state estimation error to perturbation, and to sensor noise and the sensitivity of the perturbation estimation error to perturbation, and to sensor noise.

Based on the functions, $\mathcal{S}_{\tilde{x}\eta}(z)$ in (5.16) and $\mathcal{S}_{\tilde{x}v}(z)$ in (5.17), it is possible to investigate how much the state estimation accuracy can be improved thanks to the perturbation observer. Also, based on $\mathcal{S}_{\tilde{\eta}\eta}(z)$ in (5.18) and $\mathcal{S}_{\tilde{\eta}v}(z)$ in (5.19), it can be seen how the perturbation attenuation performance of the perturbation observer (5.7) with a state estimator is changed comparing with the full-state feedback case in (5.4).

## 5.3 Application to 2nd Order Mechanical Systems

In this section, to get physical insights on the combined observer, we pay attention to 2nd order systems with single-input/single-output(SISO). Through the eigenvalue assignment design procedure and the frequency response analysis, we investigate the effect of the combined observer.

### 5.3.1 A Design Procedure

First, a pure inertia plant, $M\ddot{y} = u + w$ is considered as a nominal model of many mechanical systems, where $w$ can include even known terms as well as unknowns. If we let the state $x = [x_1 \; x_2]^T = [y \; \dot{y}]^T$, the discrete plant model (5.1) has the following matrices:

$$A = \begin{bmatrix} 1 & h \\ 0 & 1 \end{bmatrix}, \; B = \Gamma = \frac{1}{M}\begin{bmatrix} \frac{1}{2}h^2 \\ h \end{bmatrix}, \; C = \begin{bmatrix} 1 & 0 \end{bmatrix} \tag{5.20}$$

with $h$ the sampling time. Secondly, if a simple 1st order filter,

$$Q_1(z) = \frac{b}{1 - az^{-1}} \tag{5.21}$$

is considered in (5.5), the dimensions in (5.7) are $q = p = 1$ and the parameters of the perturbation observer (5.7) are given as

$$F = a, \; G = b, \; H = 1. \tag{5.22}$$

Then, with the estimator gain matrix, $L = [\ell_1 \; \ell_2]^T$, the *characteristic equation* of the combined observer error dynamics in (5.8) can be written as

$$\begin{aligned}\mathcal{D}(z) &= \det(zI - A_o) \\ &= (z-a-b)\underbrace{\left[z^2 + (\ell_1 - 2)z + 1 - \ell_1 + h\ell_2\right]}_{=\det(zI-A+LC)} + \frac{1}{2}h^2\alpha\beta b(z+1) \quad (5.23)\end{aligned}$$

where

$$\alpha = \frac{1}{2}h^2\ell_1 + h\ell_2,\ \beta = 1/(\frac{1}{4}h^4 + h^2). \tag{5.24}$$

Let the desired characteristic equation of (5.8) be

$$\mathcal{D}_d(z) = (z - \lambda_1)(z - \lambda_d)^2 \tag{5.25}$$

with all critically damped poles, where one pole is due to the perturbation observer and the other two poles are due to the Luenberger observer. Here, the DC gain of (5.21) should be one, i.e. $a + b = 1$. Then, by the eigenvalue assignment design, $\mathcal{D}(z) = \mathcal{D}_d(z)$, we have the following relationships:

$$\ell_1 = 3 - \lambda_1 - 2\lambda_d, \tag{5.26a}$$

$$\ell_2 = \frac{1}{2h}(\lambda_d - 1)\left[\lambda_1(\lambda_d + 3) + \lambda_d - 5\right], \tag{5.26b}$$

$$b = \frac{1}{h^2\alpha\beta}(1 - \lambda_1)(1 - \lambda_d)^2 = 1 - a. \tag{5.26c}$$

In the above, as the desired eigenvalue $\lambda_1 \to 1$, the estimator gains reduce to

$$\ell_1 = 2(1 - \lambda_d),\ \ell_2 = \frac{1}{h}(1 - \lambda_d)^2, \tag{5.27}$$

which is the case of pure Luenberger observer with two desired poles at $\lambda_d$.

On the other hand, $\lambda_1 \to 1$ means $b \to 0$ and $a \to 1$ in (5.26), which indicates that the cut-off frequency ($\omega_c$) of the filter (5.21) is decreased to zero. Extremely, when $\lambda_1 = 1$ ($\omega_c = 0$), the perturbation observer does not work at all in the combined observer.

### 5.3.2 Sensitivity Analysis

When applying (5.20) and (5.22) to (5.16) and (5.17), the sensitivity functions of the state ($x_1$: position, $x_2$: velocity) estimation error with respect to perturbation($w$) and sensor noise($v$) can be written respectively as

$$\mathcal{S}_{\tilde{x}w}(z) = \left\{\begin{matrix}\dfrac{\tilde{x}_1(z)}{w(z)} \\ \dfrac{\tilde{x}_2(z)}{w(z)}\end{matrix}\right\} = \left\{\begin{matrix}\dfrac{h^2}{2M}\dfrac{(z-a-b)(z+1)}{\mathcal{D}(z)} \\ \dfrac{h}{M}\dfrac{(z-a-b)(z+\ell_1-1-\frac{1}{2}h\ell_2)}{\mathcal{D}(z)}\end{matrix}\right\}, \tag{5.28}$$

$$
\mathcal{S}_{\tilde{x}v}(z) = \begin{Bmatrix} \dfrac{\tilde{x}_1(z)}{v(z)} \\ \dfrac{\tilde{x}_2(z)}{v(z)} \end{Bmatrix} = \begin{Bmatrix} -\dfrac{(z-a-b)(\ell_1 z - \ell_1 + h\ell_2) + \frac{1}{2}bh^2\alpha\beta}{\mathcal{D}(z)} \\ -\dfrac{(z-1)\left[\ell_2(z-a-b) + bh\alpha\beta\right]}{\mathcal{D}(z)} \end{Bmatrix} \tag{5.29}
$$

where $\mathcal{D}(z)$ is given in (5.23).

First, as the cut-off frequency ($\omega_c$) of the filter (5.21) goes to zero (i.e. as $b \to 0$), we have $\mathcal{D}(z) \to (z-a-b)\det(zI - A + LC)$ in (5.23) and the above functions reduce to the Luenberger observer case:

$$
\mathcal{S}_{\tilde{x}w}(z) = \begin{Bmatrix} \dfrac{\tilde{x}_1(z)}{w(z)} \\ \dfrac{\tilde{x}_2(z)}{w(z)} \end{Bmatrix} = \begin{Bmatrix} \dfrac{h^2}{2M} \dfrac{z+1}{\det(zI - A + LC)} \\ \dfrac{h}{M} \dfrac{(z + \ell_1 - 1 - \frac{1}{2}h\ell_2)}{\det(zI - A + LC)} \end{Bmatrix}, \tag{5.30}
$$

$$
\mathcal{S}_{\tilde{x}v}(z) = \begin{Bmatrix} \dfrac{\tilde{x}_1(z)}{v(z)} \\ \dfrac{\tilde{x}_2(z)}{v(z)} \end{Bmatrix} = \begin{Bmatrix} \dfrac{-(\ell_1 z - \ell_1 + h\ell_2)}{\det(zI - A + LC)} \\ \dfrac{-\ell_2(z-1)}{\det(zI - A + LC)} \end{Bmatrix}. \tag{5.31}
$$

Conversely, from (5.28) and (5.29), it can be inferred that as the cut-off frequency of (5.21) gets larger ($\omega_c \to \infty$), that is, as $a$ gets smaller or $b$ gets larger in (5.21),

i) $\|\mathcal{S}_{\tilde{x}w}(z)\|$ gets smaller at low frequencies comparing with the Luenberger observer case (5.30). That is, the state estimation error is reduced for a perturbation with allowable bandwidth, but
ii) $\|\mathcal{S}_{\tilde{x}v}(z)\|$ gets larger at high frequencies comparing with the Luenberger observer case (5.31). That is, the sensor noise is more amplified by adopting the perturbation observer in the state estimator.

The above two facts can be confirmed by Figs. 5.2 and 5.3 which denote the frequency response of (5.28) and (5.29), respectively. To generate the figures, the parameters in (5.20) were given as $M = 0.202\ V/(m/sec^2)$, $h = 1\ msec$ and the parameters in (5.26) were determined by assigning two poles at $\lambda_d = 0.5, 0.5$ and an arbitrary pole at $\lambda_1$. As shown, as the eigenvalue $\lambda_1 \to 0$, the sensitivity to perturbation in Fig. 5.2 is gradually reduced in low frequencies below the cut-off frequency($\omega_c$) of (5.21), which indicates that the state estimation accuracy is enhanced. But also, as $\lambda_1 \to 0$, the noise sensitivity in Fig. 5.3 is inversely increased in high frequencies beyond the cut-off frequency of (5.21), which means that the estimates become more noisier. This result is reasonable considering that $\lambda_1 \to 0$ means that the bandwidth of the filter (5.21) becomes wider, and therefore much of the high frequency process noises could be entered into the combined observer through the perturbation observer (5.7). Hence, in real implementations, the accuracy of estimates should be compromised considering the noise property of measurements.

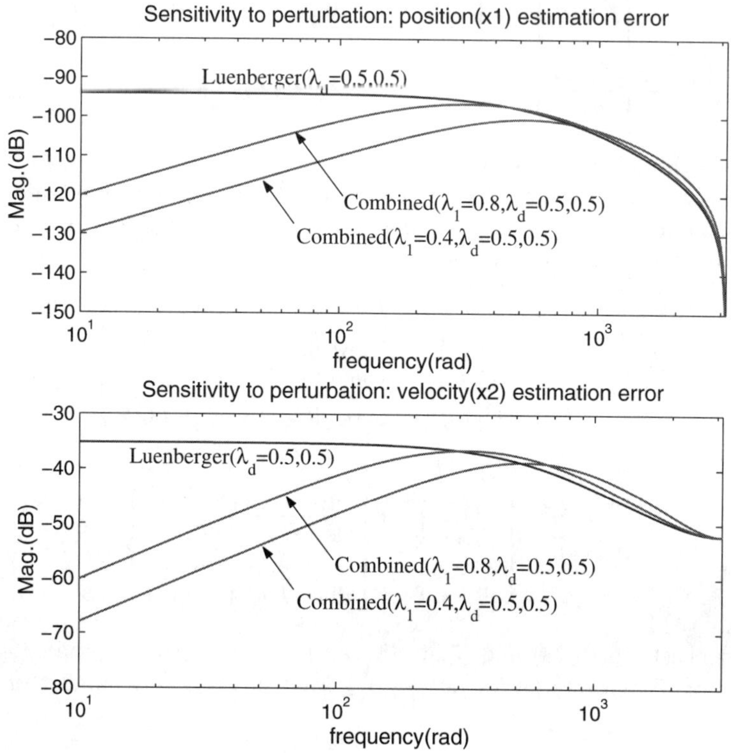


**Fig. 5.2.** Frequency response of (5.28): Sensitivity of state estimation error to perturbation.

A second issue in the combined observer is how much the performance of the perturbation observer (5.4) with full-state feedback is distorted when the state estimator is introduced as in (5.7). First, in the combined observer case, by applying (5.20) and (5.22) to (5.18), we get the sensitivity function of perturbation estimation error:

$$\mathcal{S}_{\tilde{w}w}(z) = \frac{\tilde{w}(z)}{w(z)} = \frac{(z-a-b)\left[z^2 + (\ell_1 - 2)z + 1 - \ell_1 + h\ell_2\right]}{\mathcal{D}(z)} \tag{5.32}$$

with $\mathcal{D}(z)$ in (5.23). Secondly, in full-state feedback case, by letting $w_{eq}(k) = w(k)$ and applying (5.22) to (5.4), we have $\hat{w}(z)/w(z) = b/(z-a)$. Then,

$$\frac{\tilde{w}(z)}{w(z)} = \frac{z-a-b}{z-a} = 1 - z^{-1}Q_1(z) \tag{5.33}$$

In Fig. 5.4, 'Combined' denotes the frequency response of (5.32) with desired poles at $\lambda_d = 0.5, 0.5$ and $\lambda_1$ and 'Full-state' is that of (5.33) with the same filter parameters $(a, b)$ as the 'Combined' case. As shown, the response

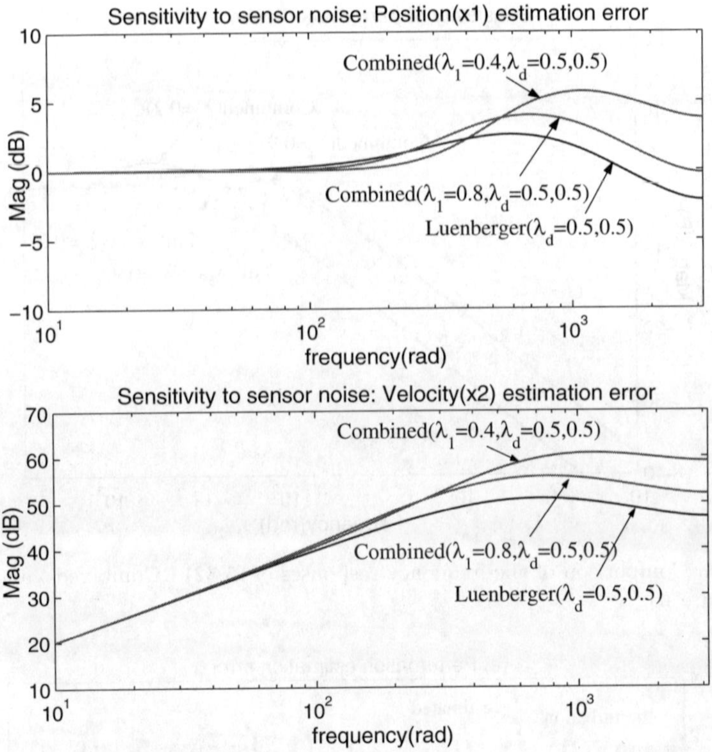


**Fig. 5.3.** Frequency response of (5.29): Sensitivity of state estimation error to sensor noise.

of (5.32) is different from that of (5.33) around the cut-off frequency($\omega_c$) of the filter (5.21), but they are almost the same in low frequencies. Moreover, according as the eigenvalue $\lambda_1$ decreases (i.e., $\omega_c$ increases), the sensitivity of the perturbation estimation error becomes smaller and the frequency band of perturbation attenuation (i.e, $\|\tilde{w}\| < \|w\|$) is enlarged.

### 5.3.3 Numerical Example

To confirm the effectiveness of the proposed combined observer, we give an example of time response simulation. Given plant is a single-axis linear positioning table whose equation of motion is described by $M\ddot{y}(t) + c\dot{y}(t) = u(t) + d(t) - C_f(\dot{y})$ with inertia $M = 0.202\ V/(m/sec^2)$, damping coefficient $c = 2.25\ V/(m/sec)$, and Coulomb friction $C_f(\dot{y}) = 0.11$(when $\dot{y} > 0$),$-0.13$(when $\dot{y} < 0$). The pure inertia model, $M\ddot{y}(t) = u(t) + w(t)$ with $(x_1 = y, x_2 = \dot{y})$ is taken as the nominal model to construct the combined observer in (5.2). Then, the matrices in the discrete state-space equation (5.1) are equal to (5.20) and the perturbation has the components of

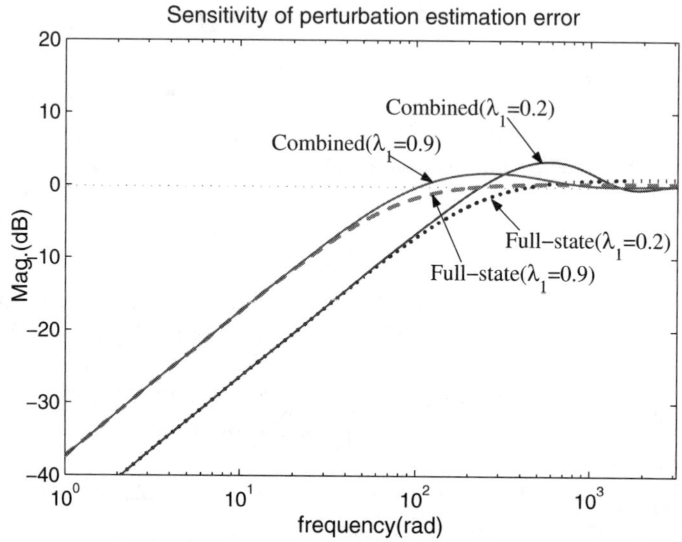


**Fig. 5.4.** Comparison of the frequency responses of (5.32) ('Combined') and (5.33) ('Full-state').

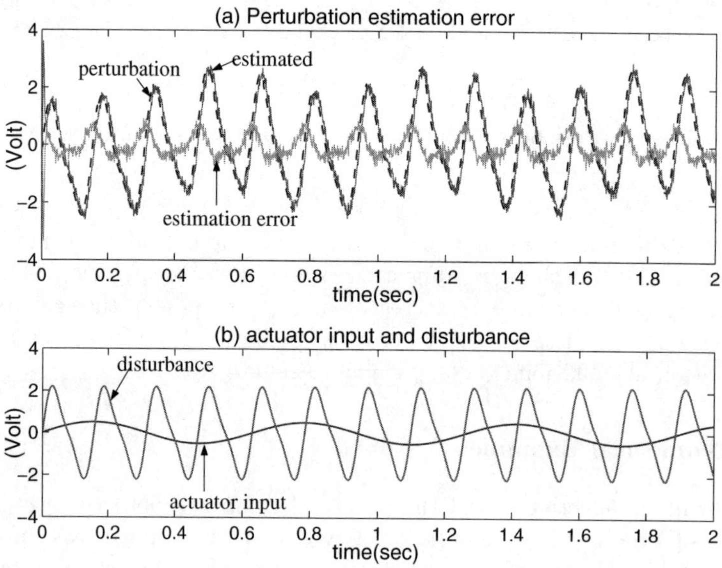


**Fig. 5.5.** (Simulation) (a) Perturbation($w(t)$) and perturbation estimation error($\tilde{w}(t)$), (b) Actuator input, $u(t)[V] = 0.5\sin(10t)$ and disturbance, $d(t)[V] = 2\sin(40t) + 0.5\sin(80t)$.

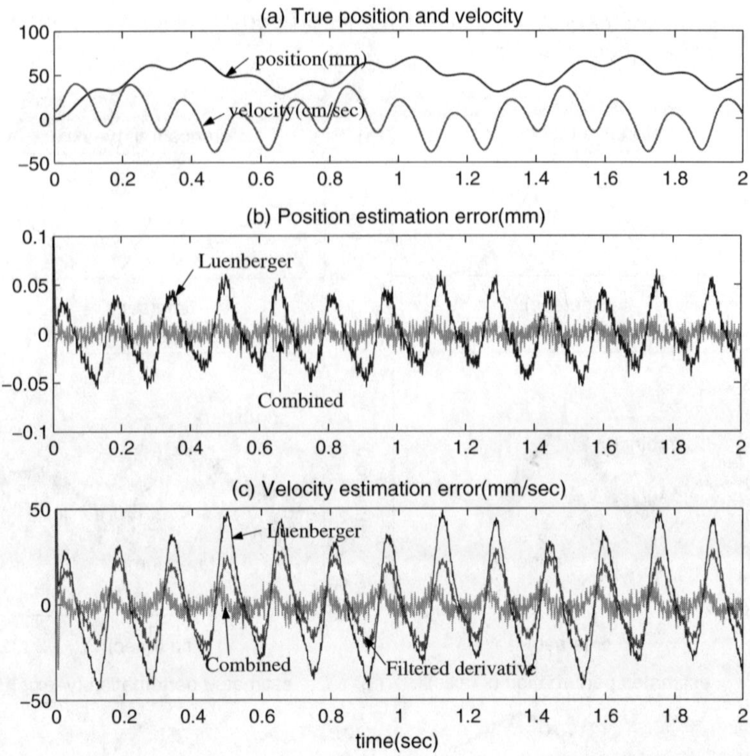


**Fig. 5.6.** (Simulation) (a) True position and velocity, (b) Position estimation error, (c) Velocity estimation error.

$w(t) = d(t) - c\dot{y}(t) - C_f(\dot{y})$. The initial condition is set as $y(0) = 0, \dot{y}(0) = 10\ cm/sec$. The sampling time is $h = 1\ msec$ and the sensor model for position measure is simply $y_m(t) = y(t) + v(t)$ where $v(t)$ is random noise with zero-mean and standard deviation $\sigma = 5 \times 10^{-6}$, i.e., $v \sim \mathcal{N}(0, \sigma^2)$. The actuator input was $u(t)[V] = 0.5\sin(10t)$ and a heavy external disturbance, $d(t)[V] = 2\sin(40t) + 0.5\sin(80t)$ was also applied.

Figures 5.6 and 5.5 are the simulation results when the estimator gains in (5.26) and (5.27) are determined for the desired poles at $\lambda_d = 0.5, 0.5$ and $\lambda_1 = 0.6$ in (5.25). As shown in Fig. 5.6(b) and 5.6(c), the state (position and velocity) estimation error was greatly reduced by the combined observer comparing with the Luenberger observer case. Figure 5.5(a) shows that the perturbation observer well estimates the real perturbation. But, in Fig. 5.6, the results of combined observer are noisier than those of Luenberger observer, which confirms the frequency response in (5.17). On the other hand, the 'Filtered derivative' in Fig. 5.6(b) is the difference between the true velocity and the velocity by numerical differentiation for the position measure,

$\dot{y}(k) = (y_m(k) - y_m(k-1))/h$ (which was filtered through a 1st order filter with cut-off frequency at 100 $Hz$).

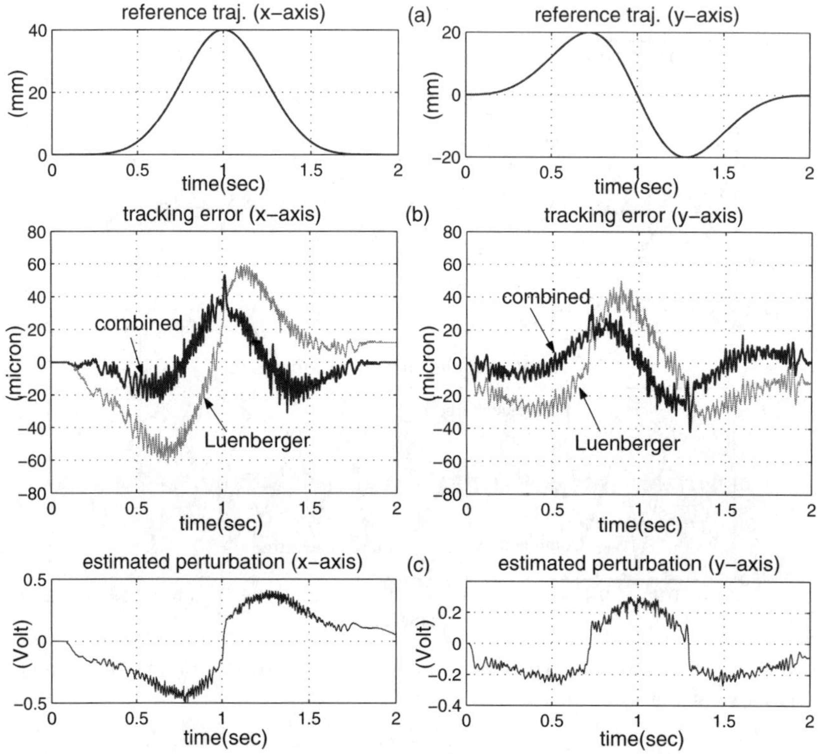


**Fig. 5.7.** (a) Reference trajectory, (b) Tracking error, (c) Perturbation estimate.

### 5.3.4 Experiment

In the above simulation results, we have seen that the performance of the combined observer is *robust* with respect to the plant perturbation. Although the accuracy of state estimate is critical in itself in diverse applications, also it is very important in control applications. In this section, the contribution of the combined observer to enhance the control performance is experimentally verified. The task is to track the circle by the XY-positioner, whose nominal model for controller-observer design is given by the pure inertia model, $M_x\ddot{x} = u_x + w_x$ for $x$-axis ($M_x = 0.202\ V/(m/sec^2)$) and $M_y\ddot{y} = u_y + w_y$ for $y$-axis ($M_y = 0.1665\ V/(m/sec^2)$). The reference trajectory of the two axes to generate circular motion with radius $R = 20\ mm$ and tracking time $t_f = 2\ sec$ are given in Fig. 5.7(a). While, the tracking controller in each axis has the

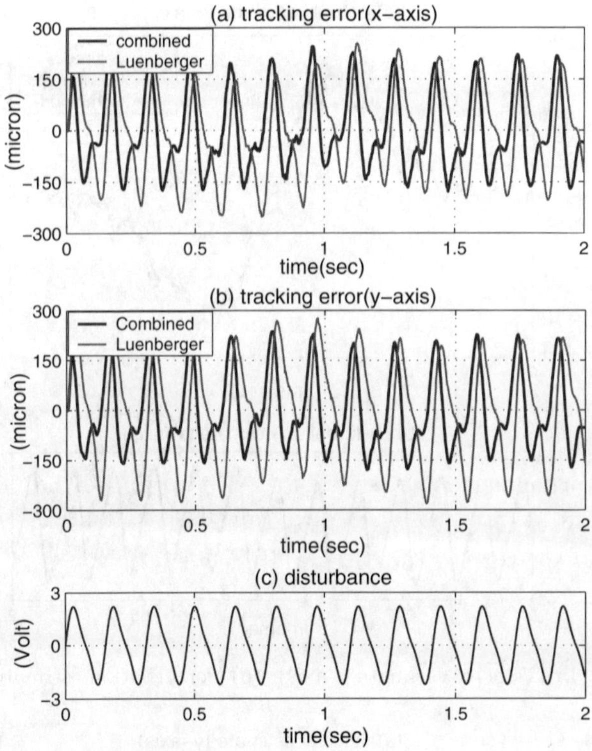


**Fig. 5.8.** (a) Tracking error(x-axis), (b) Tracking error(y-axis), (c) Input disturbance.

form of PD plus perturbation compensator, i.e., $u_k = K_P e_k + K_D \dot{e}_k - \hat{w}_k$, where $\hat{w}_k$ is given by the same form as (5.7). The position measure to get the tracking error($e_k$) is obtained from the motor encoder and the velocity error $\dot{e}_k$ is determined with the velocity estimate from the state estimator. Hence, in the tracking controller–combined observer combination, the perturbation observer contributes to both robust control and robust state estimation.

As shown in Fig. 5.7(b), the Luenberger observer case denotes biased values in final tracking errors, which are caused by the fact that the velocity estimate by the Luenberger observer is heavily distorted from the true value. This is because the uncertainty effect such as nonlinear friction cannot be suppressed effectively by the linear innovation process alone. In contrast, the combined observer case shows relatively good tracking performance. While, the perturbation estimate in Fig. 5.7(c) by the perturbation observer mainly corresponds to the viscous friction plus Coulomb friction generated during the motion of the XY-positioner in the experimental setup. Secondly, Fig. 5.8 compares the tracking performance when a heavy external disturbance in Fig. 5.8(c) is applied to both axes of the XY-positioner. As shown, the combined

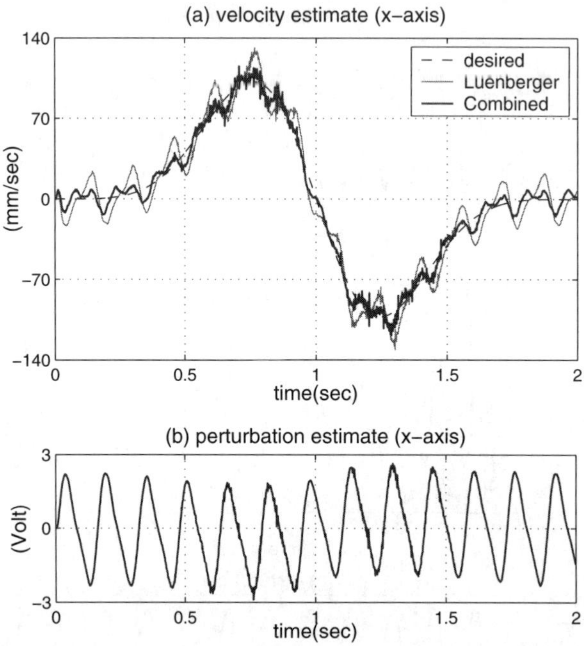


**Fig. 5.9.** (a) Velocity estimate(x-axis), (b) Perturbation estimate(x-axis).

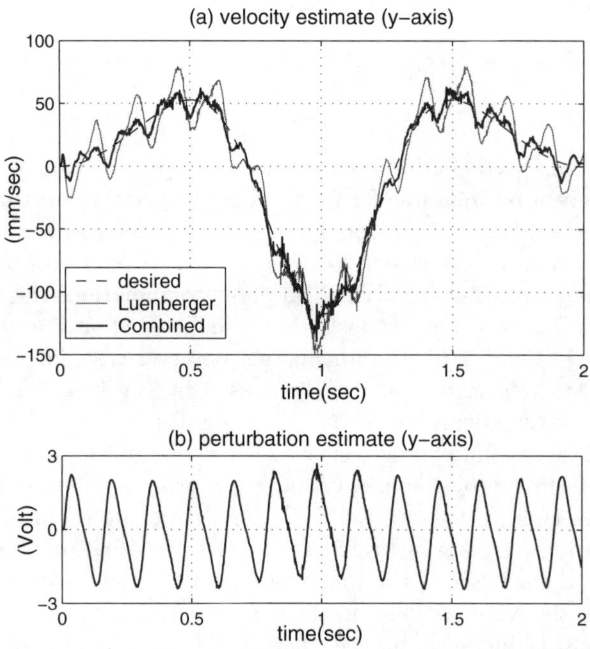


**Fig. 5.10.** (a) Velocity estimate(y-axis), (b) Perturbation estimate(y-axis).

observer produces better results. For this experiment, Figs. 5.9 and 5.10 shows that the velocity estimate by the combined observer is closer to the desired velocity profile than by the standard Luenberger observer.

## 5.4 Robust Kalman Filtering

In the former section, we investigated the combined observer synthesis for *deterministic* systems. Now, it is extended to *stochastic* systems. As is well-known, the Kalman filter [26,56] produces best linear estimates for stochastic systems given the Gaussian noise statistics of process noise and sensor noise. However, in spite of the optimality and the guaranteed stability margin, it was a controversial problem that the Kalman filter assumed a *white* noise process with zero-mean. In fact, most process noises of physical plants are not white nor do they have zero-mean value. Like the Luenberger observer in deterministic plant case, the estimate by the Kalman filter will be biased from the real state when the plant perturbation (i.e., process noise) has some bias components. This is because the standard Kalman filter intrinsically has no property to compensate the effect of such biased perturbations. In the former section, it was enabled to obtain an enhanced state estimation performance by combining the standard Luenberger observer with the perturbation observer. Likewise, the Kalman filter can be modified to fulfill this point.

First of all, let's consider a discrete plant with stochastic processes:

$$x_{k+1} = A_k x_k + B_k u_k + \Gamma_k w_k, \tag{5.34a}$$

$$y_k = C_k x_k + v_k, \tag{5.34b}$$

where the measurement noise $v_k$ is assumed to be a uncorrelated white Gaussian process with zero-mean and known covariance $R_2 \triangleq E[v_k v_k^T]$, i.e., $v_k \sim \mathcal{N}(0, R_2)$. However, considering the process noise $w_k$ (i.e., plant perturbation), we assume that it is a mixture of colored and white Gaussian process but the perturbation estimation error, $\tilde{w}_k = w_k - \hat{w}_k$ due to the perturbation observer is an uncorrelated white Gaussian process with zero-mean and known covariance $R_1 \triangleq E[\tilde{w}_k \tilde{w}_k^T]$, i.e., $\tilde{w}_k \sim \mathcal{N}(0, R_1)$.

At this point, the following notations are necessary.

$\hat{x}_{k|k-1}$ : a priori estimate before measurement update.
$\hat{x}_{k|k}$ : a posteriori estimate after measurement update.
$\tilde{x}_{k|k-1} \triangleq x_k - \hat{x}_{k|k-1}$ : a priori estimation error.
$\tilde{x}_{k|k} \triangleq x_k - \hat{x}_{k|k}$ : a posteriori estimation error.
$P_{k|k-1} \triangleq E[\tilde{x}_{k|k-1}\tilde{x}_{k|k-1}^T]$ : covariance of a priori estimation error.
$P_{k|k} \triangleq E[\tilde{x}_{k|k}\tilde{x}_{k|k}^T]$ : covariance of a posteriori estimation error.

The discrete Kalman filter has two update processes, i.e., the time update which is a state extrapolation process using the plant model and the measurement update with innovation process. In this chapter, the time update process is modified as

$$\hat{x}_{k+1|k} = A_k\hat{x}_{k|k} + B_k u_k + \Gamma_k \hat{w}_k, \tag{5.35}$$

where the perturbation observer of the last term is newly inserted. Secondly, an optimal estimate $\hat{x}_{k|k}$ which minimizes the mean-square error, $J = E[\tilde{x}_{k|k}^T \tilde{x}_{k|k}]$ is determined in the measurement update process:

$$\hat{x}_{k|k} = \hat{x}_{k|k-1} + K_k\left(y_k - C_k\hat{x}_{k|k-1}\right), \tag{5.36}$$

where the Kalman gain matrix $K_k$ can be readily derived using the *orthogonality principle* in Theorem 5.2 as follows.

$$\begin{aligned} & E[\tilde{x}_{k|k}\tilde{y}_k^T] = 0 \\ & \rightarrow E[\{\tilde{x}_{k|k-1} - K_k(C_k\tilde{x}_{k|k-1} + v_k)\}\{C_k\tilde{x}_{k|k-1} + v_k\}^T] = 0 \\ & \rightarrow K_k = P_{k|k-1}C_k^T\left(C_k P_{k|k-1}C_k^T + R_2\right)^{-1}. \end{aligned} \tag{5.37}$$

**Theorem 5.2.** *[26, 56]:* The optimal estimate of a Gaussian random process $x_k$ which minimizes the mean-square error, $J = E[\tilde{x}_k^T\tilde{x}_k] = tr(P_k)$ is the conditional expectation of $x_k$ given the measurements $y_k$, i.e., $\hat{x}_k = E[x_k|y_k]$. Moreover, in case of Gaussian process, the optimal estimate is *linear* with respect to measurements, i.e., $\hat{x}_k = \alpha_1 y_k + \alpha_0$ and it is equal to the projection of $x_k$ onto $\mathcal{Y}(k)$ the linear vector space of measurements, that is, the estimation error $\tilde{x}_k$ is *orthogonal* to $\mathcal{Y}(k)$. □

Since a posteriori estimate $\hat{x}_{k|k}$ is closer to the real state than the a priori estimate $\hat{x}_{k|k-1}$, it is desirable to use the a posteriori estimate in the perturbation observer (5.7). And if 1st order filters,

$$Q_i(z) = \frac{b_i}{1 - a_i z^{-1}}, \quad i = 1, \cdots, p \tag{5.38}$$

are used in (5.7), we have $F = \text{diag}\{a_1, \cdots, a_p\}$, $G = \text{diag}\{b_1, \cdots, b_p\}$, $H = I_p$, and $F + G = I_p$ where $p = \dim(w_k)$. Then, the perturbation estimate update process can be written as

$$\hat{w}_k = F\hat{w}_{k-1} + G\Gamma_{k-1}^{+}\left(\hat{x}_{k|k} - A_{k-1}\hat{x}_{k-1|k-1} - B_{k-1}u_{k-1}\right). \tag{5.39}$$

To update the Kalman gain (5.37) in real-time, two error covariance update processes, $P_{k|k-1}$ and $P_{k|k}$ are needed. First, from (5.34) and (5.35), the following error dynamics is given:

$$\tilde{x}_{k|k-1} = A_{k-1}\tilde{x}_{k-1|k-1} + \Gamma_{k-1}\tilde{w}_{k-1}, \tag{5.40}$$

where it should be noted that $\tilde{w}_k$ and $\tilde{x}_{k|k}$ are mutually uncorrelated. Then, by substituting (5.40) to $P_{k|k-1} \triangleq E[\tilde{x}_{k|k-1}\tilde{x}_{k|k-1}^T]$ and noting that $\tilde{w}_k \sim$

$\mathcal{N}(0, R_1)$ is assumed, we obtain the propagation equation of the a priori error covariance:

$$P_{k|k-1} = A_{k-1}P_{k-1|k-1}A^T + \Gamma_{k-1}R_1\Gamma_{k-1}^T. \tag{5.41}$$

Second, by subtracting both sides of (5.36) from $x_k$ and using the measurement model in (5.34), we have

$$\tilde{x}_{k|k} = \tilde{x}_{k|k-1} - K_k\left(C_k\tilde{x}_{k|k-1} + v_k\right). \tag{5.42}$$

Then, by substituting (5.42) into $P_{k|k} \triangleq E[\tilde{x}_{k|k}\tilde{x}_{k|k}^T]$, noting $v_k \sim \mathcal{N}(0, R_2)$, and using the relationship of (5.37), the covariance equation of the a posteriori error is given as

$$P_{k|k} = (I - K_kC_k)\,P_{k|k-1}. \tag{5.43}$$

By arranging the above results, we have the recursive estimation formula in Table 5.1, where the underlined parts are newly inserted processes when comparing with the standard discrete Kalman filter. On the other hand, by substituting (5.42) into (5.40), a predictive estimation error dynamics is given by

$$\tilde{x}_{k+1|k} = A_k\left(I - K_kC_k\right)\tilde{x}_{k|k-1} - A_kK_kv_k + \Gamma_k\tilde{w}_k, \tag{5.44}$$

where the last exciting term $\tilde{w}_k$ will contain those harmonic components which are too fast to be detected by the perturbation estimation process in (5.39).

*Remark 5.3.* By the perturbation estimation process in (5.39), the colored part of process noise whose frequency spectrum is below the cut-off frequency of $Q_i$ filters in (5.38) can be well estimated. In Table 5.1, the perturbation estimate recursively affects the time update process. In this way, the plant perturbation will be copied to the Kalman filter. As a result, we have the error dynamics in (5.44) with $\|\tilde{w}_k\| < \|w_k\|$, which indicates that the accuracy of the state estimate could be greatly enhanced by reducing the norm $\|\tilde{x}_k\|$ compared with the case of not applying the perturbation estimation process. □

*Remark 5.4.* Regrading the innovation process in (5.36), it is a kind of simple proportional control to regulate the state estimate to the real state. In contrast, the perturbation estimation process in (5.39) has an adaptive characteristic for the time-varying perturbation. □

### 5.4.1 Numerical Example

Here, we take a numerical example which shows the contribution of the perturbation observer to the robust Kalman filtering. The simulation model is identical to that in the former section 5.3.3. But now, the control input is set

**Table 5.1.** Discrete Kalman filter with perturbation observer.

- Time update:
  $$\hat{x}_{k|k-1} = A_{k-1}\hat{x}_{k-1|k-1} + B_{k-1}u_{k-1} + \underline{\Gamma_{k-1}\hat{w}_{k-1}}$$
- A priori error covariance update:
  $$P_{k|k-1} = A_{k-1}P_{k-1|k-1}A_{k-1}^T + \Gamma_{k-1}R_1\Gamma_{k-1}^T$$
- Kalman gain update:
  $$K_k = P_{k|k-1}C_k^T\left(C_kP_{k|k-1}C_k^T + R_2\right)^{-1}$$
- Measurement update:
  $$\hat{x}_{k|k} = \hat{x}_{k|k-1} + K_k\left(y_k - C_k\hat{x}_{k|k-1}\right)$$
- A posteriori error covariance update:
  $$P_{k|k} = \left(I - K_kC_k\right)P_{k|k-1}$$
- Perturbation estimate update:
  $$\underline{\hat{w}_k = F\hat{w}_{k-1} + G\Gamma_{k-1}^{+}\left(\hat{x}_{k|k} - A_{k-1}\hat{x}_{k-1|k-1} - B_{k-1}u_{k-1}\right)}$$

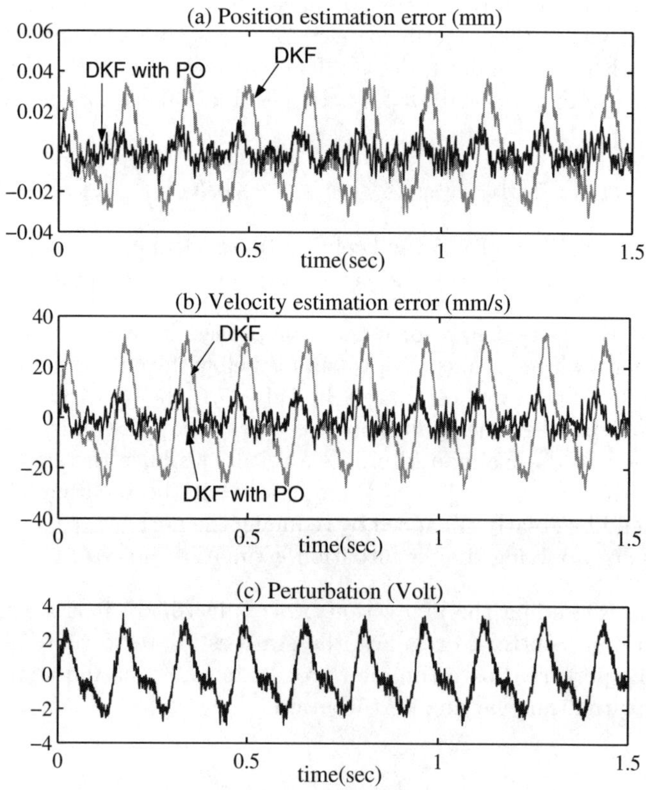


**Fig. 5.11.** Performance of the discrete Kalman filter(DKF) with and without the perturbation observer(PO): (a) position estimation error, (b) velocity estimation error, (c) process noise (perturbation).

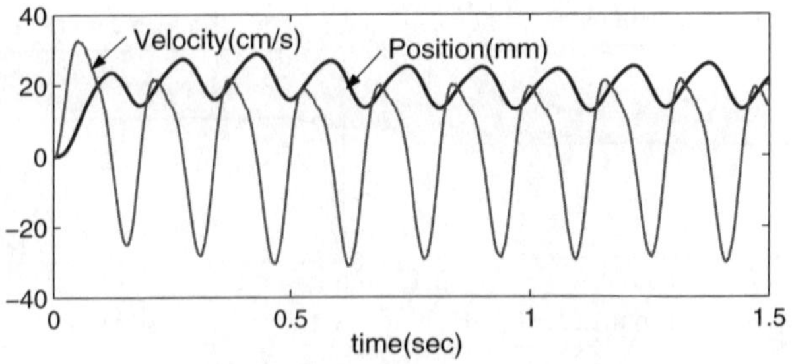


**Fig. 5.12.** Position and velocity profile.

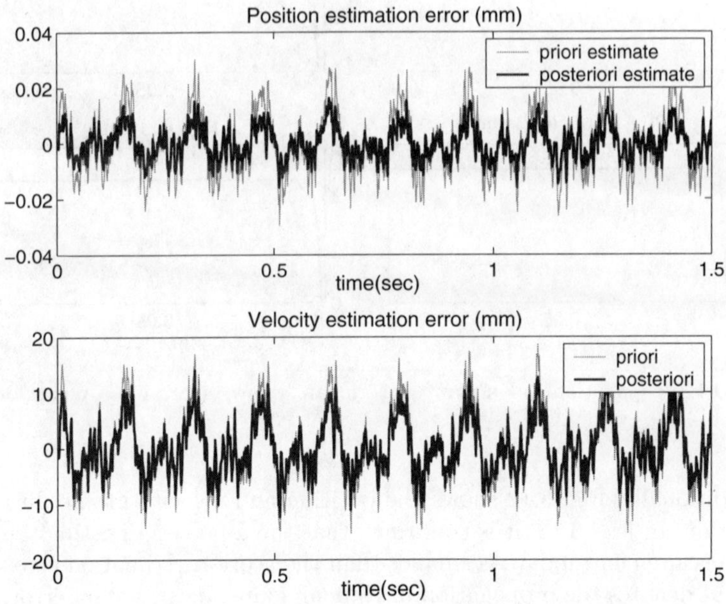


**Fig. 5.13.** Comparison of a priori estimation error and a posteriori estimation error.

to zero ($u_k = 0$) and the process noise (plant perturbation) is assumed to be the sum of the external disturbance, $d_k = 2\sin(40kh) + \sin(80kh)$, friction, and the random noise $\xi_k \sim \mathcal{N}(0, R_1)$, i.e., $w_k = d_k - c\dot{y}(t) - C_f(\dot{y}) + \xi_k$, which is denoted in Fig. 5.11(c). When the update processes in Table 5.1 are applied with the covariances $R_1 = 0.3^2$ and $R_2 = (5\times10^{-6})^2$ and the low-pass-filtering parameters in (5.39) are given as $F = a = 0.7$ and $G = b = 0.3$, we have the results in Fig. 5.11(a) and (b), where the performances of the Kalman filter with and without the perturbation estimate process are compared with one another. As shown, the state estimation error is greatly reduced with the action of the perturbation estimate update process.

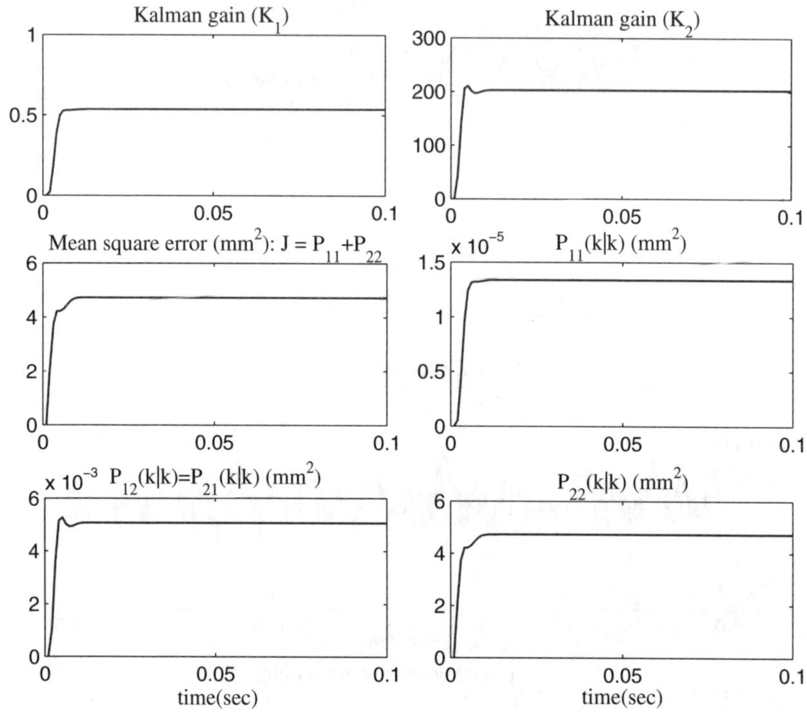


**Fig. 5.14.** Propagation of Kalman gains, mean square error, and a posteriori error covariance.

Additionally, Fig. 5.12 shows the position and velocity profile during the simulation. In Fig. 5.13, it is confirmed that the a posteriori estimation error after measurement update is smaller than the a priori estimation error. Also, Fig. 5.14 denotes the propagation of Kalman gains, mean square error, and a posteriori error covariance, $P_{k|k} = \begin{bmatrix} P_{11} & P_{12} \\ P_{21} & P_{22} \end{bmatrix}$, respectively.

## 5.5 Summary

In this chapter, a robust state estimation method was investigated through the combined synthesis of state estimator and perturbation observer. The combined observer is characterized by the recursive algorithm where the state estimator and the perturbation observer are dynamically inter-connected, where the perturbation estimation process enables to acquire reliable state estimates in spite of heavy external disturbances. Following the derivation of the coupled estimation error dynamics of plant states and perturbation, four kinds of sensitivity functions were determined. In application to 2nd order mechanical systems, the sensitivity analysis addresses the perturbation observer's contribution to the accuracy of state estimate and also the limit due to the sensor noise. The combined observer setting was also extended to stochastic systems and the discrete Kalman filter was reformulated by involving the perturbation estimate update process. The numerical example and experimental results confirmed the validity of the combined observer.

From a practical implementation point of view, the combined observer of this chapter is promising since it requires no knowledge of plant uncertainty above all. Based on the results of this chapter, it is not wise to use the Luenberger observer or the Kalman filter in their standard forms, specifically when the plant is under heavy perturbation. In the case of mechanical systems, if not adopting the perturbation observer, at least the Coulomb friction model should be inserted into the state estimator to get reliable estimates.

# 6

# Control of Coarse/Fine Dual-Stage Positioning System

This chapter, although it is a little bit off the subject of this book, is included to show an application example where the perturbation compensator contributes to control system performance.

## 6.1 Introduction

In mechanical systems driven by conventional *coarse actuators* such as electrical motors or hydraulic actuators, the servo bandwidth and the resolution of motion are limited mainly by the stick/slip motion due to the nonlinear friction around the zero velocity and the resonance mode at high frequency band. On the other hand, *fine actuators* such as piezoelectric actuators are free of friction and capable of following high frequency commands, but their travel ranges are extremely small. Hence, as a trial to increase the servo bandwidth and the motion resolution of conventional positioning systems, a compound actuation system with coarse/fine (or macro/micro) dual-stage architecture which adopts the merits of both coarse and fine actuator could be a quite attractive approach. In dual-stage systems, the coarse actuator provides large power and long travel range while the fine actuator enables fast and fine motions.

Taking some examples where the concept of coarse/fine compound actuation was used, first, some researchers paid attention to the macro/micro architecture for robot manipulators [30, 40, 83], where the micro (or mini) manipulator corrects the positioning error due to gear backlash or structural flexibility in the conventional manipulator. Secondly, the interest on the dual-stage actuation was very active recently in the data storage systems such as hard disk drives [29, 57, 79, 103], where the micro actuators usually made by micro fabrication technology were mounted on the suspension actuated by the voice coil motor(VCM) and performs high speed track following over the VCM bandwidth. Also, there were several works which adopted dual-stage linear positioner for robotic applications [49, 50, 53, 58].

The former approaches for dual-stage control are mainly classified into multi-input and multi-output (MIMO) design and single-input and single-output (SISO) design according to the plant characteristics. Since the dual-stage positioner is a multi-input/multi-output system, the MIMO design can be considered like the LQG/LTR design [103] and the $\mu$-synthesis [29]. However, when the dynamic interaction between the coarse and fine stage is negligible, the SISO design is sufficient. In addition, the dual-stage servo architecture can be changed according to sensory condition. For example, in dual-stage hard disk drives, i) when the position error signal(PES) at the dual-stage end-point and also the relative position error signal(RPES) of the fine actuator are available, a dual-input/dual-output type controller [57] can be designed, but ii) when only the PES is available, a single-input/dual-output type controller such as [79] is inevitable. Here, the basics for the dual-stage servo design can be stated as follows.

- The desired trajectory should be tracked by the coarse stage with long travel range while the fine stage with high servo bandwidth corrects high frequency tracking errors resulting from the bandwidth limit of the coarse actuator.
- From the viewpoint of frequency loop shaping, the frequency response of the coarse stage controller must have a large gain in the low frequency region so that the coarse stage operates primarily in response to low frequency inputs. To the contrary, the fine stage controller must have relatively a large gain in the high frequency region so as to compensate high frequency perturbations.

In this chapter, two control problems which were not so deeply discussed in other dual-stage articles will be mainly investigated. First, owing to the small travel range, fine stage motions may be easily saturated for some large reference inputs. Hence, to make the most of the fine actuator capability of fast and fine motion, the dual-stage controller must have the function of avoiding the range saturation. To fulfill this point, we suggest a perturbation compensator to maintain small tracking error of the coarse stage and determine a null-motion control input to possibly reduce the relative distance between the two stages. Second, another benefit of the dual-stage system lies in the fact that it can increase the minimum-time control performance in comparison with single-stage case. Related to this, we propose a practical minimum-time control strategy in terms of predetermined time-optimal trajectory and robust perturbation compensator.

This chapter is organized as follows. First of all, a control framework for dual-stage system is suggested in Section 2, where are given the perturbation compensator for robust tracking control and the null-motion control input which defines the internal motion between the coarse and fine stage. In Section 3, a minimum-time control problem using the coarse/fine dual-stage is discussed. In Section 4, we show an example of micro-teleoperation which adopts the dual-stage as a slave manipulator. Finally, conclusion is followed in Section 6.

## 6.2 Dual-Stage Controller Design

### 6.2.1 Physical Characteristics

To understand the advantage of dual-stage compound actuation, a physical property of dual-stage systems is considered. According to [40], the operational space kinetic energy matrix of a manipulator can be described as $\mathbf{\Lambda}(\mathbf{x}) = \mathbf{J}^{-T}(\mathbf{q})\mathbf{A}(\mathbf{q})\mathbf{J}^{-1}(\mathbf{q})$, where $\mathbf{A}(\mathbf{q})$ is the manipulator inertia matrix, and $\mathbf{J}(\mathbf{q})$ is the Jacobian between the joint velocity($\dot{\mathbf{q}}$) and the end-effector velocity($\dot{\mathbf{x}}$) in operational space. Moreover, the effective inertia, $\sigma_w(\mathbf{\Lambda})$ perceived at the end-effector along a specified direction $\mathbf{w}$ is given by $\sigma_w(\mathbf{\Lambda}) = \frac{1}{\mathbf{w}^T\mathbf{\Lambda}^{-1}\mathbf{w}}$. Then, in manipulators with macro/micro(i.e., coarse/fine) architecture, the effective inertia satisfies the following relationship.

$$\sigma_w(\mathbf{\Lambda}_o) \leq \sigma_w(\mathbf{\Lambda}_m) \tag{6.1}$$

in the direction of motion $\mathbf{w}$, where $\mathbf{\Lambda}_o$ and $\mathbf{\Lambda}_m$ are the operational space kinetic energy matrices respectively associated with the overall macro/micro manipulator and the single micro manipulator.

At this point, let's consider the linear dual-stage system in Fig. 6.1 where two stages are colinearly mounted. It is assumed that the bottom and top stage with respective inertia $H_1$ and $H_2$ correspond to the coarse and fine stage, respectively. Then, the joint coordinates are $\{y_1, y_2\}$, the operational space coordinate is $y$, the Jacobian is $\mathbf{J} = [1 \ \ 1]$, the inertia matrix is $\mathbf{A} = \mathrm{diag}\{H_1, H_2\}$, and $\mathbf{\Lambda}_m$ in (6.1) is equal to $H_2$. Applying the above equation (6.1) to this system, the effective inertia, $H_{end}$ at the end-point of the dual-stage along the operational direction $y$ is readily derived as

$$H_{end} = \frac{H_1 H_2}{H_1 + H_2} \leq H_2. \tag{6.2}$$

That is, the effective inertia of the dual-stage becomes smaller than that of the single-stage. This means that the dual-stage system can produce larger end-point acceleration than single-stage cases. This result is reasonable considering the fact that the bottom stage adds power to the base of the top stage in Fig. 6.1.

Although the dual-stage system gives a chance to enhance the positioning performance, it may be accompanied by some adverse problems. Above all, in mounting the second stage, the overall off-axis error (i.e., runout from the ideal straight line) may become worse than the single-stage case. Hence, when only built-in joint sensors such as encoder are used for control, a strict kinematic calibration through external measurements is required to numerically correct

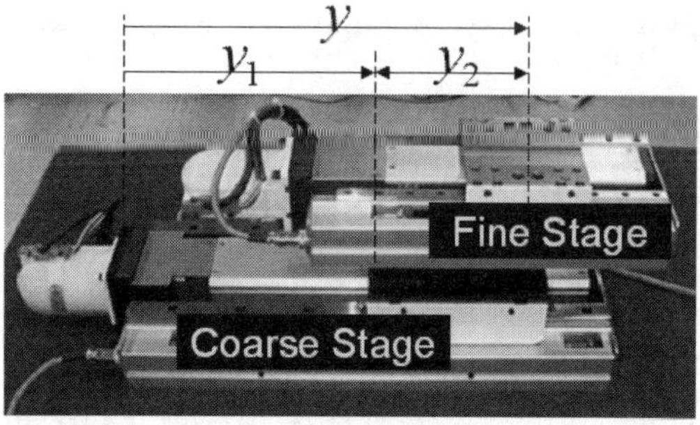


**Fig. 6.1.** An example of dual-stage positioner.

such mechanical errors. However, if some external sensors are used for end-point feedback, this problem can be naturally relaxed. As shown in Fig. 6.1, since the dual-stage architecture which mounts the second stage in the same direction has the actuation redundancy, an infinite number of position sets are possible for a given end-point position. Hence, the reference inputs to both stages should be properly resolved considering the capability and the physical constraint of each actuator.

### 6.2.2 A Dual-Stage Control Architecture

Figure 6.2 denotes the dual-stage control architecture proposed in this chapter. As shown, the reference generator produces the desired trajectory $y_d$ of the end-point, the controller $C_1$ makes the coarse stage follow the desired trajectory $y_d$, i.e., $y_1 \to y_d$, and the fine stage controller $C_2$ corrects the end-point error. In Fig. 6.2, the fine stage controller $C_2$ requires the position error $e$ at the end-point. When only local positions $(y_1, y_2)$ are measured, it can be determined as $e = e_1 - y_2$. (note that $e = y_d - y = y_d - (y_1 + y_2) = e_1 - y_2$.) In other words, the coarse stage tracking error $e_1$ is the reference input to the fine stage. However, if the absolute position of the end-point can be measured, we obtain directly $e = y_d - y$ and the feedback loop of the fine stage in Fig. 6.2 can be accordingly altered.

Since the travel range of most fine actuators is extremely small (usually, under hundreds of micron in piezoelectric actuators), the fine stage motion could be easily saturated to its travel limit for some large reference commands beyond the fine stage range. During the saturation interval, the fine stage loses its capability to compensate high frequency disturbances or to follow high frequency commands. Moreover, it may occur wind-up phenomena under integral control. Hence, the dual-stage controller must have a function to avoid the situation that the saturation is maintained for a long time. In the linear dual-stage positioner in Fig. 6.1, let the travel range of the top fine stage $|y_2| \leq R$. Then, the movable range of the fine stage to correct the coarse stage tracking error $e_1$ is defined as $S(y_2) = R - \text{sgn}(e_1)y_2$ at the current position

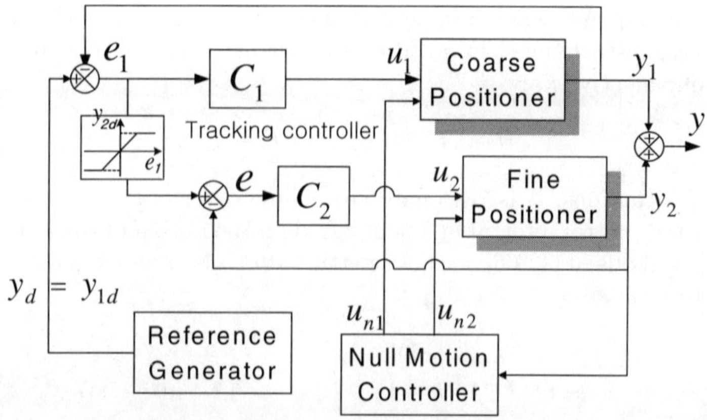


**Fig. 6.2.** Dual-stage control architecture.

$y_2$. The saturation block in Fig. 6.2 restricts the reference input to the fine stage as

$$y_{2d} = \begin{cases} e_1 & \text{for } |e_1| < S \\ \text{sgn}(e_1)S & \text{for } |e_1| \geq S \end{cases} \tag{6.3}$$

so that it does not exceed the movable range, $S(y_2)$.

To prevent the range saturation of the fine stage, above all, the tracking error of the coarse stage should be maintained as small as possible. The perturbation compensator designed in the next section enables robust tracking performance in spite of heavy perturbation to the plant. In Fig. 6.2, when $e_1 \to 0$ by $C_1$ and $e \to 0$ by $C_2$, naturally $y_2 \to 0$. However, the null-motion controller investigated in the following section accelerates the action of $y_2 \to 0$ using the internal motion of the dual-stage and much reduces the possibility of fine stage saturation.

### 6.2.3 Perturbation Compensator for Robust Tracking Control

The control performance of coarse actuators is degraded primarily by the nonlinear friction due to the hard contact of moving parts. Also, the positioning accuracy of fine actuators(e.g., piezoelectric actuators) can be largely distorted due to the hysteresis phenomena. Hence, to get robust tracking performance in spite of the time-varying perturbations, it is desirable to include an additional perturbation compensation logic in the tracking controller as well as normal feedback control law.

In [48, 50], we proposed an improved perturbation compensator. Here, it is briefly described. First, we consider the mass-damper system

$$H_n\ddot{y}(t) + B_n\dot{y}(t) = u(t) + \psi(t) \tag{6.4}$$

as the nominal plant, where the perturbation $\psi(t)$ includes model uncertainty and external disturbances. From (6.4), the perturbation to the nominal plant can be equivalently expressed as

$$\psi_{eq}(t) = H_n \ddot{y}(t) + B_n \dot{y}(t) - u(t). \tag{6.5}$$

In normal situations, it is reasonable to assume $\psi(t) \approx \psi(t-h)$ during the discrete-time control interval $h$. Then, based on the notion of time-delayed estimation methods in [33,66], a causal perturbation observer can be determined in discrete form as

$$\hat{\psi}(k) = Q \cdot \psi_{eq}(k-1) \tag{6.6a}$$
$$= Q \cdot \left( H_n \ddot{y}(k-1) + B_n \dot{y}(k-1) - u(k-1) \right), \tag{6.6b}$$

where the low pass filtering operator $Q$ is to attenuate sensor noise and other high frequency harmonics.

On the other hand, by using the reference trajectory $(y_d, \dot{y}_d, \ddot{y}_d)$, we can obtain a feedforward-type perturbation observer as

$$\hat{\psi}_f(k) = Q \cdot \psi_{eq}(k-1, y_d) \tag{6.7a}$$
$$= Q \cdot \left( H_n \ddot{y}_d(k-1) + B_n \dot{y}_d(k-1) - u(k-1) \right). \tag{6.7b}$$

If (6.7) is applied to the plant (6.4), a large amount of perturbation can be compensated. However, there remains a residue, $\psi_r(t) = \psi(t) - \hat{\psi}_f(t) = \psi_{eq}(t) - \hat{\psi}_f(t)$ still not attenuated by (6.7). A secondary observer to further compensate the residual perturbation is given as

$$\hat{\psi}_r(k) = Q \cdot \psi_r(k-1) = Q \cdot \left( \psi_{eq}(k-1) - \hat{\psi}_f(k-1) \right) \tag{6.8}$$

By combining (6.7) and (6.8), an improved perturbation compensator which uses both feedback and feedforward signals can be obtained:

$$\hat{\psi}_c(k) = \hat{\psi}_f(k) + \hat{\psi}_r(k). \tag{6.9}$$

Then, the robust control input is defined as the sum of nominal feedback control input $\mu(k)$ and the perturbation compensation input $\hat{\psi}_c(k)$, i.e., $u(k) = \mu(k) - \hat{\psi}_c(k)$. In Fig. 6.2, the tracking controller can be considered to include the above perturbation compensator.

**Experimental Example:** Figure 6.3 is the experimental result for the bottom ball-screw stage in Fig. 6.1 to compare the performance of the perturbation compensator (6.9) with (6.6). In both cases, the same nominal feedback control law was applied, the same low pass filter $Q$ was used in the perturbation compensators (6.6), (6.7), and (6.8), and an arbitrary perturbation $\psi(t) = 3\sin(20t)[Volt]$ was added to the control command. As shown, the improved perturbation compensator (6.9) enables better tracking performance.

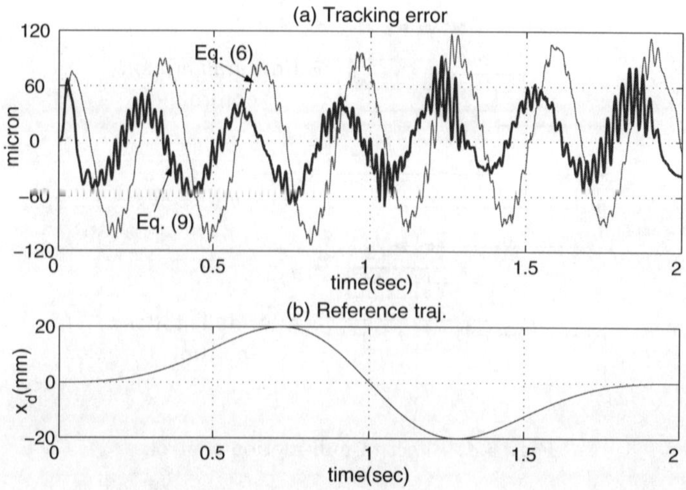


**Fig. 6.3.** (a) Performance comparison of (6.6) and (6.9). (b) Reference trajectory.

### 6.2.4 Null-Motion Controller

In dual-stage systems, due to the extra degrees of freedom, there exist null-motions (i.e., internal motions) between the two stages which do not produce end-point motions. By using the null-motion capability, another set of control inputs can be determined to minimally keep the distance of the fine stage from the middle point($y_2 = 0$) without perturbing the end-point position. It helps to extend the movable range of fine stage during the dual-stage control for the same end-point tracking error. Figure 6.4 denotes the concept of the null-motion control. That is, the relative motion($y_2$) between the two stages can be quite reduced while the end-point position($y$) is not perturbed.

In fact, the null-motion control is a well-established topic for robot manipulators with kinematic redundancy. In redundant manipulators, the relationship between the end-effector force($\mathbf{F}$) and the joint torques($\mathbf{\Gamma}$) takes the following form:

$$\mathbf{\Gamma} = \mathbf{J}^T\mathbf{F} + \left[\mathbf{I} - \mathbf{J}^T\mathbf{J}^{T+}\right]\mathbf{\Gamma}_o, \tag{6.10}$$

where $\mathbf{J}$ is the Jacobian matrix. The second term in (6.10) is the null-space joint torques, where $\mathbf{J}^{T+}$ is the generalized inverse of $\mathbf{J}^T$ and $\mathbf{\Gamma}_o$ is an arbitrary generalized joint torque vector which will be projected onto the null space of $\mathbf{J}^T$. According to [40], a generalized inverse of $\mathbf{J}^T$ with which the null-space joint torques do not produce any operational acceleration at the end-effector(i.e, which satisfies the so-called dynamic consistency) is given by the inertia weighted pseudo-inverse:

$$\bar{\mathbf{J}}(\mathbf{q}) = \mathbf{A}^{-1}\mathbf{J}^T(\mathbf{J}\mathbf{A}^{-1}\mathbf{J}^T)^{-1} \tag{6.11}$$

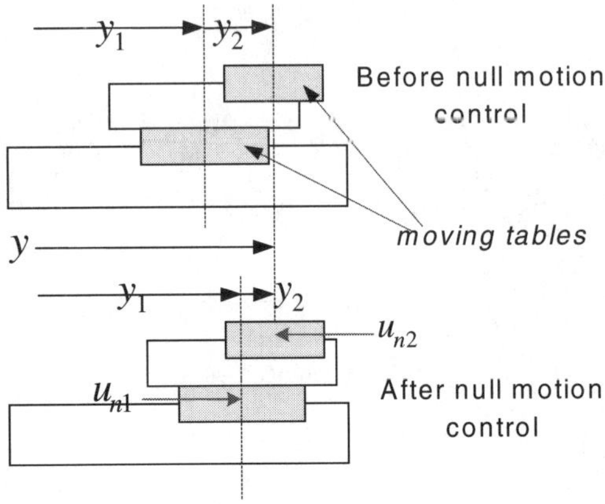


**Fig. 6.4.** Effect of null-motion control.

where **A** is the manipulator inertia matrix. Accordingly, the null-motion control inputs which do not dynamically affect the end-effector motion is given as

$$\mathbf{\Gamma}_n = \left[\mathbf{I} - \mathbf{J}^T \bar{\mathbf{J}}^T\right] \mathbf{\Gamma}_o. \tag{6.12}$$

Applying the above equation to the dual-stage system in Fig. 6.1 with $\mathbf{J} = [1 \;\; 1]$ and $\mathbf{A} = \text{diag}\{H_1, H_2\}$, the null-motion control input is readily derived as follows.

$$\mathbf{\Gamma}_n = \begin{Bmatrix} u_{n1} \\ u_{n2} \end{Bmatrix} = \begin{bmatrix} -\frac{1}{4}\left(\frac{H_1+H_2}{H_2}\right) \\ 1 - \frac{1}{4}\left(\frac{H_1+H_2}{H_2}\right) \end{bmatrix} \mathbf{\Gamma}_o \tag{6.13}$$

Hence, for example, if the generalized joint torque is given by

$$\mathbf{\Gamma}_o = -K_P y_2 - K_D \dot{y}_2 \tag{6.14}$$

with proper PD gains, the input (6.13) will attract the fine stage to the middle point, i.e., $y_2 \to 0$ while not perturbing the end-point motion. As a result, with the null-motion control, the possibility of fine stage saturation can be considerably reduced and rather large end-point errors can be compensated by the fine stage.

**Experimental Example:** The proposed dual-stage control loop in Fig. 6.2 was applied to the dual-stage positioner in Fig. 6.1. The tracking controller has the PD rule with the perturbation compensator. Although both the bottom and the top stage were driven by electrical motors, coarse/fine dual-stage motion was mimicked by adjusting control gains so that the top stage has five times higher servo bandwidth than the bottom stage. Figure 6.5 shows experimental results for the end-point reference trajectory in Fig. 6.5(c). As shown, the coarse stage tracking error ($e_1$) was effectively compensated by the fine stage motion ($y_2$) and the end-point has small tracking error($e$). Considering the effect of the null-motion control, the relative motion of the fine stage to the coarse stage was much reduced with the null-motion control (Fig. 6.5(b)) in comparison with the case without the null-motion control (Fig. 6.5(a)), but the end-point error was almost not changed. This validates the null-motion control effect. In the experiment, although no friction compensation scheme was applied, the coarse stage showed almost zero steady state error thanks to the perturbation compensator.

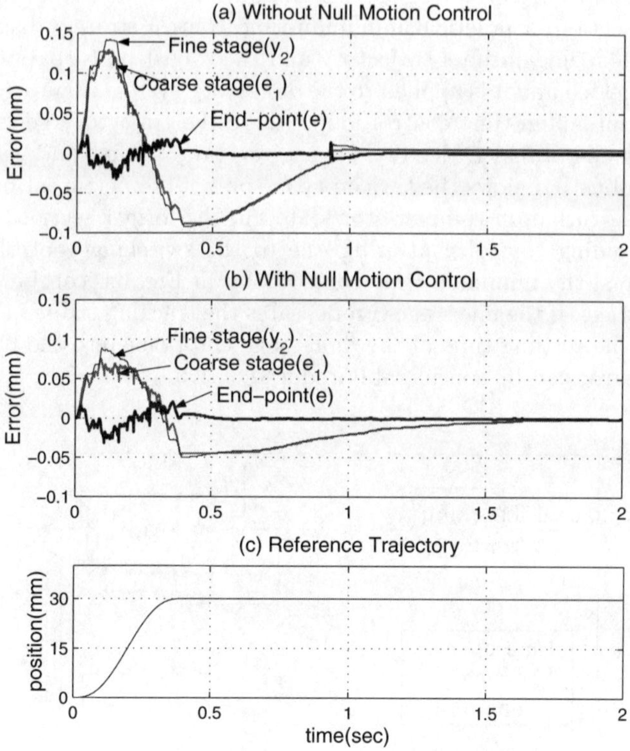


**Fig. 6.5.** Dual-stage control experiment. (a) Without null-motion control. (b) With null-motion control. (c) Reference trajectory.

## 6.3 Minimum-Time Control with Coarse/Fine Dual-Stage

The minimum-time control is considered as a critical problem in many applications, for example, in track seeking mode of hard disk drives [23, 61] and in pick and place work of fast chip mounting devices. Since the dual-stage positioner enables higher servo bandwidth than conventional positioner, the time-optimal performance can be accordingly enhanced. For example, a time-optimal seek trajectory was investigated for dual-stage disk drives in [61], where the idea to reduce the seek time was to make the coarse actuator overshoot the target point as much as the moving range of the fine actuator. Usually, time-optimal control law is given by bang-bang type feedback law [7]. However, the bang-bang control is not practical since it has no margin on the model uncertainty and it may occur heavy chattering owing to switching control actions. Moreover, the minimum-time control performance much depends on the settling time to the target point in addition to the rising time.

In this section, a practical minimum-time control strategy based on the predetermined time-optimal trajectory and the robust perturbation compensator is suggested and it is applied to the dual-stage system. Figure 6.6 denotes the overall minimum-time control loop, where the trajectory generator produces the time-optimal trajectory for a given target point and the tracking controller robustly follows the trajectory in spite of the perturbation with the aid of the perturbation compensator stated in the former section. It has the merit of avoiding control chattering due to the switching control. In dual-stage systems, the minimum-time control loop in Fig. 6.6 can be applied to the coarse stage. If the fine stage compensates the tracking errors in the coarse stage well, the settling time of the dual-stage end-point into the tolerance of regulation error can be considerably reduced.

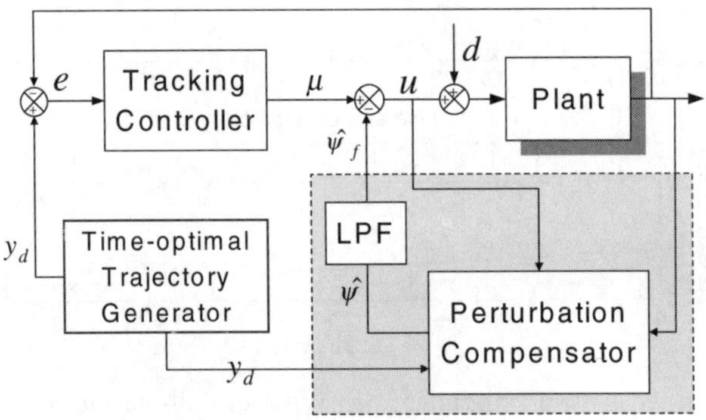


**Fig. 6.6.** A robust minimum-time control loop.

### 6.3.1 Time-Optimal Trajectory

The time-optimal control problem is to find a control which minimizes the cost function, $J = \int_{t_0}^{t_f} dt$ with the constraint of state equation, $\dot{x} = f(x, u, t)$ and limited actuation power $|u| \leq u_{max}$. In terms of the Pontrygin's minimum principle [7], the time-optimal control law is given by the bang-bang control:

$$\nu(t) = -\text{sgn}(h(x)) = \begin{cases} -1 & \text{if } h(x) > 0 \\ +1 & \text{if } h(x) < 0 \end{cases} \tag{6.15}$$

where $h(x)$ is the switching function and $\nu$ is the normalized input($|\nu| \leq 1$).

In fact, the heart of time-optimal control problem is to determine the switching function $h(x)$. However, the solution was known for a few specific linear plants. For example, for a pure inertia plant, $\ddot{y} = b\nu$, the switching function is simply given as $h(x) = y + (1/2b)\dot{y}|\dot{y}|$, $x = \{y, \dot{y}\}$ for an arbitrary initial condition, $\{y, \dot{y}\}(t_0)$ and the zero target condition, $\{y, \dot{y}\}(t_f) = \{0, 0\}$. Although the pure inertia plant is assumed to be the nominal model for time-optimal control of many positioning systems [23], it is inadequate in high speed systems because the portion of friction force becomes so large according to the velocity increase. Instead, the mass-damper plant model, $H_n\ddot{y} + B_n\dot{y} = u$ including linear viscous friction is more practical for high speed systems. It can be rewritten as $\ddot{y} + a\dot{y} = b\nu$, where $|\nu| \leq 1$ and $|u| \leq u_{max}$. Then, applying the minimum principle, we can readily derive the following switching function for the zero target condition, $\{y, \dot{y}\}(t_f) = \{0, 0\}$.

$$h(x) = y(t) - \text{sgn}(\dot{y})\frac{b}{a^2}\ln\left\{1 + \frac{a}{b}|\dot{y}|\right\} + \frac{1}{a}\dot{y}(t). \tag{6.16}$$

Then, by using the switching function (6.16), the following time-optimal trajectory can be derived for the zero initial condition, $\{y, \dot{y}\}(t(0)) = \{0, 0\}$ and the target condition, $\{y, \dot{y}\}(t_f) = \{y_f, 0\}$.

1) When $t \leq t_s$ (switching time): acceleration interval ($\nu = +1$),

$$t_s = \ln\left\{1 - \sqrt{1 - \exp\{-(a^2/b)y_f\}}\right\}, \tag{6.17a}$$

$$y(t) = (b/a)t + (b/a^2)\left(\exp(-at) - 1\right), \tag{6.17b}$$

$$\dot{y}(t) = (b/a)\left(1 - \exp(-at)\right), \tag{6.17c}$$

$$y(\dot{y}) = -(b/a^2)\ln\{1 - (a/b)|\dot{y}|\} - (1/a)\dot{y}(t). \tag{6.17d}$$

2) When $t_s < t \leq t_f$ (targeting time): deceleration interval ($\nu = -1$),

$$t_f = t_s + (1/a)\ln\{1 + (a/b)\dot{y}(t_s)\}, \tag{6.18a}$$

$$y(t) = y(t_s) - (b/a)(t - t_s) \tag{6.18b}$$
$$+(1/a)\left(\dot{y}(t_s) + (b/a)\right)\left[1 - \exp(-a(t - t_s))\right],$$

$$\dot{y}(t) = \dot{y}(t_s)e^{-a(t-t_s)} - (b/a)\left[1 - \exp(-a(t - t_s))\right], \tag{6.18c}$$

$$y(\dot{y}) = y_f + (b/a^2)\ln\{1 + (a/b)|\dot{y}|\} - (1/a)\dot{y}(t). \tag{6.18d}$$

Figures 6.7 and 6.8 show the time-optimal trajectory (TOT) generated for the plant model of the bottom stage in Fig. 6.1: $\ddot{y} + 11.14\dot{y} = 5.94(\nu - 0.024)$ ($H_n = 0.202V/(m/s^2), B_n = 2.25V/(m/s), u_{max} = 5V$) and the target position $y_f = 30mm$, where it was compared with the 5th order polynomial trajectory (5PT) commonly used and the velocity curve of 5PT denotes the maximum which does not exceed that of TOT. As shown, the velocity profile of TOT is asymmetric (while that of 5PT is symmetric), which is caused by the reflection of the viscous friction in determining TOT. The larger the task range is, the more prominent the asymmetry is. While, the position curve of 5PT has the following form:

$$y(t) = c_0 + c_1 t + c_2 t^2 + c_3 t^3 + c_4 t^4 + c_5 t^5. \tag{6.19}$$

If the initial/target condition is given as $\{y, \dot{y}\}(t_0) = (0, 0)$ and $\{y, \dot{y}\}(t_f) = (y_f, 0)$ respectively, the coefficients of (6.19) are determined as $c_0 = c_1 = c_2 = 0$, $c_3 = 10y_f/t_f^3$, $c_4 = -15y_f/t_f^4$, and $c_5 = 6y_f/t_f^5$.

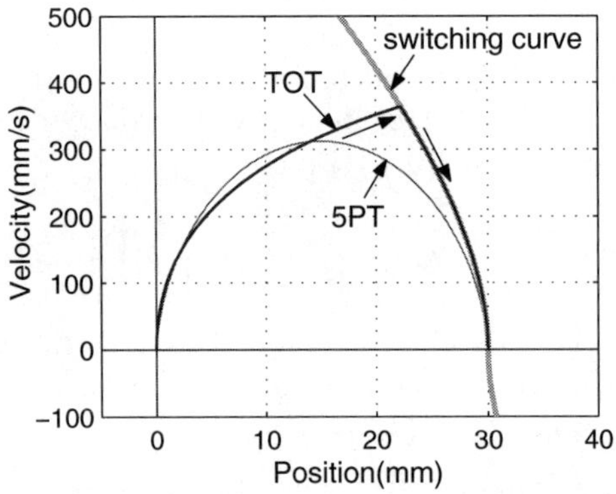


**Fig. 6.7.** Comparison of TOT and 5PT in phase plane.

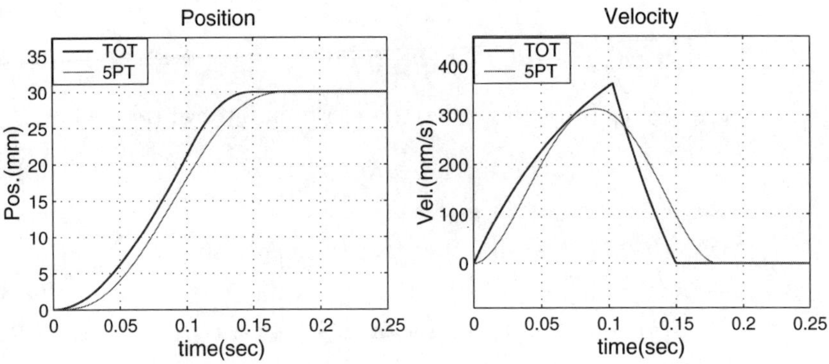


**Fig. 6.8.** Comparison of TOT and 5PT: position and velocity.

### 6.3.2 Experimental Example

The dual-stage control loop in Fig. 6.2 was applied to the $x$-direction($x = x_1 + x_2$) of the dual-stage XY positioner in Fig. 6.10. The time-optimal trajectory in (6.17)-(6.18) was determined for the coarse stage model parameters, the target point $y_f = 30\ mm$, and the limited input $u_{max} = 1.2Volt$. The tracking controller $C_1$ for the coarse stage is composed of the PD rule and the perturbation compensator in the former section but a simple PI controller was applied to the built-in piezoelectric actuator in the fine stage. An experimental result is shown in Fig. 6.9. Comparing the tracking error of the coarse stage (Fig. 6.9(a)) with that of the end-point (Fig. 6.9(c)), the settling time to the target point has been fairly reduced at the end-point($x$) by the fast and fine correction motions($x_2$) of the fine stage. This example validates the effectiveness of the suggested minimum-time control strategy using coarse/fine dual-stage.

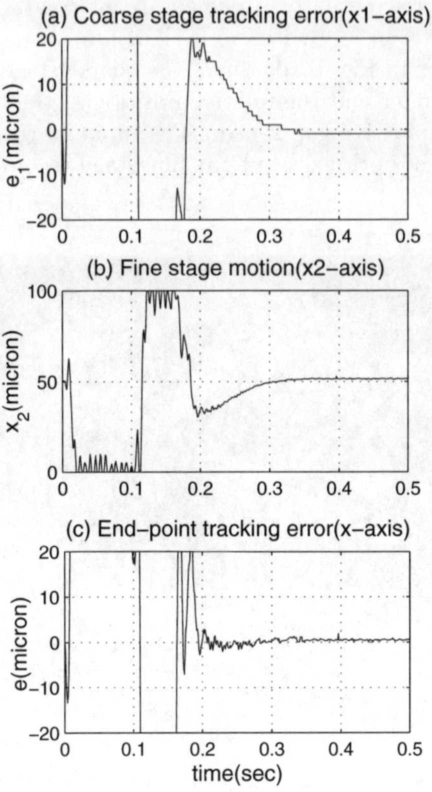


**Fig. 6.9.** Minimum-time control experiment using coarse/fine dual-stage. (a) Coarse stage tracking error. (b) Fine stage motion. (c) End-point tracking error.

## 6.4 Application to Micro-Teleoperation

In recent years, micro-teleoperation is one of the most active research area in robotics field. In the micro-teleoperation system to manipulate microscopic objects, the slave system which determines the position of the microgripper must have the function of positioning capability with high resolution (at least sub-micron) and proper workspace. The coarse/fine dual-stage discussed in the above sections is needed to fulfill this point.[1] Figure 6.10 shows the dual-stage XY positioner with 3-DOF micromanipulator constructed to use as a slave system in micro-teleoperations, where the dual-stage follows $xy$ plane motions of the master arm and the micromanipulator provides angular motions to determine the posture of the microgripper. In the dual-stage positioner, the bottom coarse XY stage has large workspace of $200 \times 100\ mm$ in $(x_1, y_1)$ plane and it is driven by BLDC motors of which rotary motions are converted into linear motions by ball-screw and linear guide. In contrast, the top fine stage has small travel range of $100 \times 100\ \mu m$ in $(x_2, y_2)$ plane and it is driven by built-in piezoelectric actuators. The end-point coordinates are defined as $x = x_1 + x_2$ and $y = y_1 - y_2$. Figure 6.11 denotes the control loop for the dual-stage positioner in Fig. 6.10, where the coarse stage position $(x_1, y_1)$ is detected by the encoder and the relative position of the fine stage $(x_2, y_2)$ is measured by the capacitive gap sensor. The input range of the coarse stage is $u = -5 \sim 5\ Volt$ in both axes but the fine stage receives unipolar input of $u = 0 \sim 10\ Volt$

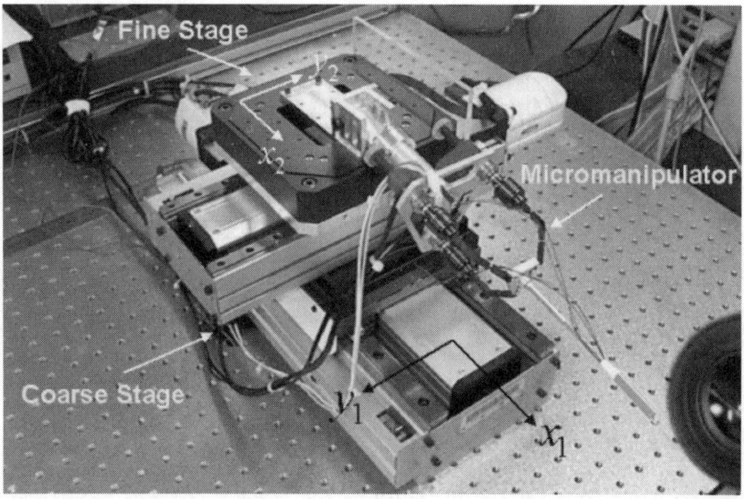


**Fig. 6.10.** Coarse/fine dual-stage XY positioner (with 3 DOF micromanipulator for posture control).

[1]See Appendix for the details of the micro-teleoperation system.

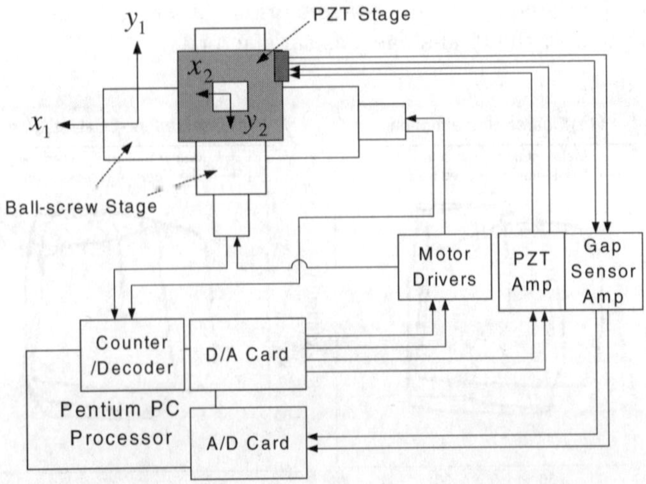


**Fig. 6.11.** Control loop of the dual-stage positioner.

Since the ball-screw driven stage in Fig. 6.10 is structurally very stiff and the dynamic coupling effect between the two axes is very weak, the motion of each axis can be described by the simple mass-damper model as $H_n\ddot{y} + B_n\dot{y} = u$. The nominal parameters were identified as the effective inertia $H_n[Volt/(m/s^2)] = \text{diag}\{0.2020,\ 0.1665\}$ and viscous damping coefficient $B_n[Volt/(m/s)] = \text{diag}\{2.25,\ 1.35\}$. While, fine stage motions are generated by the expansion/contraction of built-in piezoelectric stack actuators with the elastic deformation of flexure hinges internally equipped to guide the linear motion. Hence, it is free of friction and even nano resolution is possible. However, piezoelectric actuators denote highly nonlinear hysteresis phenomena, i.e., the output depends on the input history. So, it is very difficult to design a model-based controller. Although some researchers investigated mathematical models for piezoelectric actuators [1, 11], a simple proportional-integral (PI) controller was applied to the fine stage in this chapter.

Figure A.3 in the Appendix shows the experimental setup for micro-teleoperation, which is composed of master arm (haptic interface), microscope, slave manipulator, micro gripper, etc. The dual-stage control loop in Fig. 6.2 was applied to both axes of the dual-stage XY positioner. To achieve microscopic motions in the slave manipulator, the macroscopic motion of the master arm by the human operator were scaled down as much as 1/200. Figure 6.12 denotes the trajectory of the coarse stage[2] and that of the dual-stage in Fig. 6.10, respectively during 25 seconds motion of the master arm. Figure 6.13 shows the time history of tracking errors. Comparing Figs. 6.12(a) and 6.12(b), the coarse stage produced considerably large tracking errors but it

[2]The ball-screw driven XY stage in the dual-stage in Fig. 6.10

was well compensated by the fine stage motion. As a result, the tracking error at the end point of the dual-stage was satisfactory.

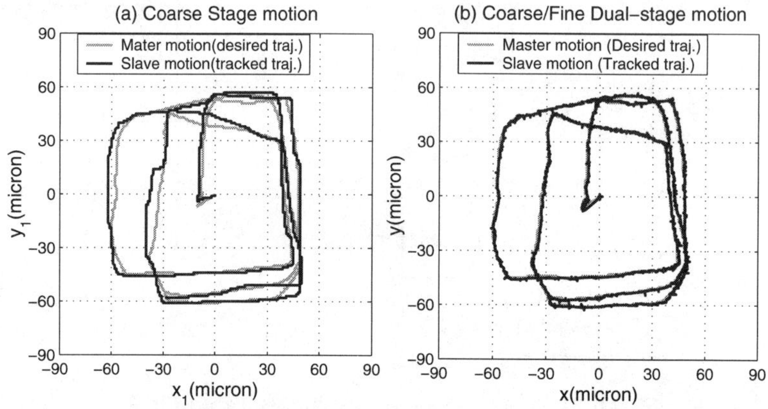


**Fig. 6.12.** (Micro-teleoperation) Tracking error (a) at the coarse stage, (b) at the dual-stage end-point.

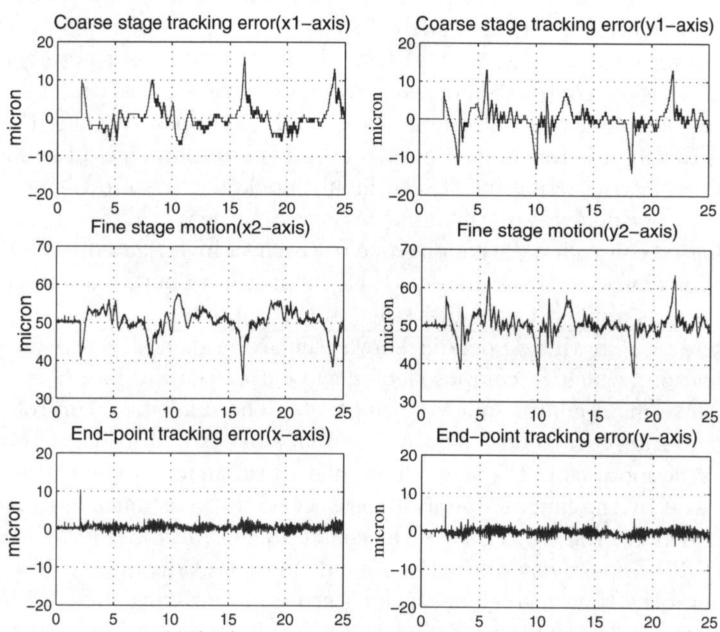


**Fig. 6.13.** (Micro-teleoperation) Coarse stage tracking error, Fine stage motion, and End-point tracking error.

### 6.4.1 Fine Trajectory Tracking

Figure 6.14 shows an example of fine contour tracking in 2-D $xy$-plane using the dual-stage XY positioner in Fig. 6.10, where the dual-stage control loop in Fig. 6.2 was applied to both $(x, y)$ axes. The reference contour is the infinitesimal circle with radius of $R = 100\ \mu m$. The corresponding reference trajectory of both axes are $x_d(t) = R(1 - cos\theta(t))$, $y_d(t) = Rsin\theta(t))$, respectively, where $\theta(t)$ is given as a 5th order polynomial trajectory for the target angle, $\theta_f = 2\pi$ and targeting time, $t_f = 2sec$. As shown in Fig. 6.14, the peaks in coarse stage tracking errors are well compensated by the fine stage motions(the travel range is $\pm 50\ \mu m$ about the neutral position 50 $\mu m$) and the tracked contour at the end-point is so clean.

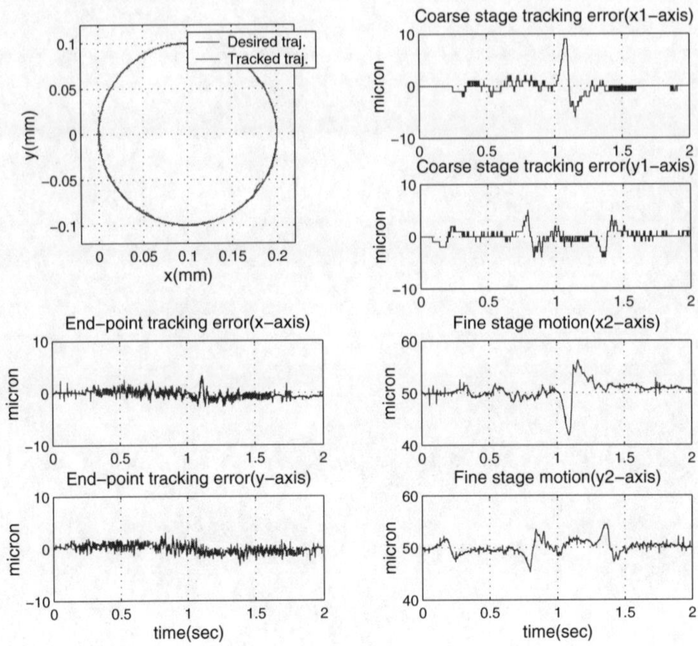


**Fig. 6.14.** Fine contour tracking using the dual-stage positioner: Tracked contour (circle of 100 $\mu m$ radius), End-point tracking error, Coarse stage tracking error, and Fine stage motion.

## 6.5 Summary

In this chapter, we considered the coarse/fine dual-stage as a solution to overcome the limit of conventional actuators. A dual-stage control architecture was proposed to effectively utilize the capability of the dual-stage system. The main points were the perturbation compensator design for robust tracking control and the null-motion controller design to make use of the extra degree of freedom in the dual-stage. In addition, a minimum-time control strategy using the coarse/fine dual-stage was suggested. The advantages of the dual-stage mechanism with the proposed algorithm were shown through experimental examples including micro-teleoperation.

# 7

# Concluding Remarks

The popularity of PID control in industry is due to its plant model independence in hardware implementation and in controller design in many cases. In the PID control process, the tuning of integral gain will be critical to well attenuate the effect of external disturbances. As a matter of fact, every controller must have an integral control property to tackle plant uncertainties. We can take an example that the integral state is augmented to the state-space plant model even in optimal control designs [14, 73]. This is to give the integral function to the resulting controller so that the closed-loop system shows asymptotic stability for constant disturbances.

The robust tracking controller in this book finally takes the form of PD plus perturbation observer. As remarked in chapter 2, the perturbation observer actually has an integral control property. Therefore, the integral function in the PID control is no more necessary as far as the perturbation observer is applied. But the superior robust performance of the perturbation observer to the simple PID control is resulted from its adaptive control property for time-varying perturbations. In fact, as the perturbation of plant becomes severer, it cannot be well suppressed by the simple integral control process with fixed gain. Instead, a direct compensation method is advantageous to meet the performance robustness. The perturbation observer in this book can be regarded as a kind of model-based dynamic compensator where the plant perturbation is estimated indirectly using just nominal plant model.

As we have seen in the former chapters, the design procedure of perturbation observer is very simple and physically intuitive. In the sequel, the form of PD control plus perturbation observer can be recommended as a promising alternative to the PID control for many industrial control systems. While, the topics covered in this book, robust motion control and robust state estimation of mechanical systems, can be categorized as follows.

**Design:** The novel idea for perturbation observer design was the hierarchical perturbation compensation method based on the notion of residual perturbation. As a consequence, we developed the hierarchical perturbation compensator (HPC) in Chapter 2, the multi-loop perturbation compensator (MPEC) in Chapter 3, and the discrete composite perturbation observer in Chapter 4. These schemes enabled to upgrade the performance of conventional perturbation observers. In Chapter 5, the strategy for robust state estimator design was to combine the perturbation observer with Luenberger observer for deterministic systems and Kalman filter for stochastic systems.

**Analysis:** The clue to investigate the property of perturbation observer was first to derive the perturbation compensation error dynamics as a function of plant modeling error (inertia uncertainty) and perturbation observer parameters ($Q$-filter). Using the transfer function and stability condition of the error dynamics, it was examined in what manner the performance and the stability robustness of perturbation observer are changed depending on $Q$-filter parameters, digital control frequency, and the number of compensation loops. Specifically, the results in Chapter 4 well addresses how the perturbation observer works in discrete-time control systems. The sensitivity analysis for the combined observer in Chapter 5 showed how it differs from the standard state observer appeared in usual text books.

**Application:** The developed perturbation compensators and the robust state estimator were successfully implemented in real mechanical systems and the advantages of proposed methods were fully demonstrated throughout the experimental results in each chapter. The coarse/fine dual-stage system in Chapter 6 was a little bit off the topic but the suggested dual-stage controller included perturbation-observer-based tracking controller which contributed to the fine positioning of microgripper in the micro-teleoperation.

Finally, some topics to be further investigated in relation to the perturbation observer approach are described in the following.

**1. Application of perturbation observer to CNC machining center.**

The quality of products in CNC machining center is naturally dependent upon the performance of servo controller. To accomplish fine tool positioning, two problems must be solved first. It includes the compensation of the frictional force generated in the XY work table and the cutting force between the tool and workpiece. Generally, the CNC servo controller has the function of model-based friction compensation. As well, the CNC machine usually provides a velocity control option to adjust feed drive velocity according to the magnitude of cutting force. Hopefully, the hierarchical perturbation compensator in this book can make the CNC machine intelligently accommodate to the time-varying nature of friction and cutting force. Another primary concern in the CNC machine control is to reduce the contour error rather than to reduce the tracking error in each axis. It is well known that the decrease of the tracking error in a axis do not necessarily mean the reduction of the contour error. To achieve good contour tracking performance, above all, time-synchronization between different axes is very important. There exists a class of contour control methods [12, 46, 47, 102] called cross-coupled controller (CCC). However, this makes the closed-loop dynamics of independent axes coupled one another. Then, we can think of designing a perturbation compensator to attenuate the effect of coupled dynamics due to the cross-coupled controller. While, since the servo controller of commercial CNC is usually not of open architecture, the hardware interface for experimental setup should be solved to implement the hierarchical perturbation compensator.

**2. Theoretical mixing of perturbation observer with optimal control.**
As a robust control framework, a defect of the class of perturbation observers such as disturbance observers [39,44,45,54,71,97], time delay estimation techniques [34,37,64,66,106], and others is that no quantitative robustness measure is addressed in these schemes. On the contrary, apart from the practicality of the 2-norm and the $\infty$-norm, the optimal control frameworks such as LQG/LTR control, $H_\infty$ control, and $\mu$-synthesis [16,17,67] clearly address the robustness issue given the state-space equation of plant, uncertainty model, and weighting functions. On the other hand, in [18, 19, 48, 66, 69, 90], it was shown that the perturbation observer greatly relaxes the conservativeness of the sliding controller by reducing the fixed switching gain for the same uncertainty bound. Similarly, if the perturbation observer is used together with optimal controller, the optimal control gain to guarantee the stability robustness for the given uncertainty bound can be much reduced. Then, an academic issue given to us is to find a complete mathematical solution on the stability and optimality for the combined optimal controller–perturbation observer.

**3. Contribution of perturbation observer in bilateral teleoperation.**
In the micro-teleoperation in Chapter 6, the role of perturbation observer was confined to compensate the uncertainty effect in the coarse/fine dual-stage. However, the perturbation observer could also contribute to compensate the other uncertainties in overall teleoperation system. A critical problem in force-reflecting micro-teleoperation is that the human operator realistically feels the microscopic gripping force transferred from the remote side of slave manipulator through the haptic interface. This problem is related to the realization of impedance transparency between the operator and the remote environment. But it is well known that the transparency conflicts with the stability of the teleoperation network [52]. To guarantee the robustness for the desired transparency and stability, the controller should be structured so that the uncertainty such as time-delay effect in communication lines as well as the model uncertainty of local master/slave system is properly addressed. In this respect, it is possible to utilize the perturbation observer to regulate the dynamics of overall teleoperation system to the desired nominal dynamics.

# A

# Appendix

## A.1 Identification of XY Table

Throughout the chapters, the XY table in Fig. 2.9 was used as a test bed to verify proposed control algorithms and to confirm corresponding analysis results. Here, it is explained how the XY table parameters were identified.

To well describe the rigid body behavior of mechanical systems, the minimal parameters to be identified are inertia, damping coefficient, and spring constant (stiffness coefficient) for individual motion axes. The linear XY table in Fig. 2.9 is a ball-screw driven type, where the angular motion of AC servo motor is converted to linear motion through the ball-screw/ball-nut mechanism and the linear motion is guided by the precision LM guide. Because the structural stiffness of the XY table is extremely high, the spring force almost do not contribute to the stage motion and it can be neglected in the plant model with no problem. And there are almost no inertial and frictional cross couplings between the two perpendicular axes. Hence, the linear mass-damper system is acceptable for the plant model of each axis.

Without depending on the measurement equipment such as logic analyzer which can generate frequency response of plant dynamics, the inertia and the damping coefficient of each axis can be simply identified in the following manner. First, obtain the "input voltage versus velocity" curves in Figs. A.1 and A.2 by applying constant voltage input and recording the resulting steady state velocity for $X$ and $Y$-axis respectively. From the graphs, we can determine the damping coefficient $B_n$ (slope of the graph) and the Coulomb friction ($C_n$, static friction at zero velocity). Second, compensate the frictional force ($B_n\dot{z} + C_n$) by feedback with P-control input. By applying the control input $u(t) = -K_P\dot{z} + B_n\dot{z} + C_n$ for each axis, the XY table will show the following dynamics.

$$\begin{aligned} H\ddot{z} + B\dot{z} &= u(t) - C = -K_P\dot{z} + B_n\dot{z} + C_n - C \\ \rightarrow\ H\ddot{z} + K_P\dot{z} &= (B_n - B)\dot{z} + C_n - C \simeq 0, \end{aligned}$$

**Table A.1.** Nominal values of XY table parameters.

| | X-axis | Y-axis |
|---|---|---|
| $H_n[Volt/(m/s^2)]$ | 0.2020 | 0.1665 |
| $B_n[V/(m/s)]$ | 2.25 | 1.35 |
| $C_n(+)[Volt]$ | 0.11 | 0.09 |
| $C_n(-)[Volt]$ | -0.13 | -0.29 |

where $H$ is the real inertia, and $B$ and $C$ are the real damping coefficient and the real Coulomb friction, respectively. Third, tune the frictional coefficients ($B_n$, $C_n$) until the time series denotes a pure sinusoidal oscillation. From the time response graph, find the period ($T$)of oscillation. Then, finally, the inertia can be determined by

$$\omega_n^2 = \left(\frac{2\pi}{T}\right)^2 = \frac{K_P}{H_n} \quad \rightarrow \quad H_n = K_P \frac{T^2}{4\pi^2}$$

In Table A.1 with the identified parameter values, $C_n(+)$ and $C_n(-)$ denote the Coulomb friction in positive and negative direction, respectively.

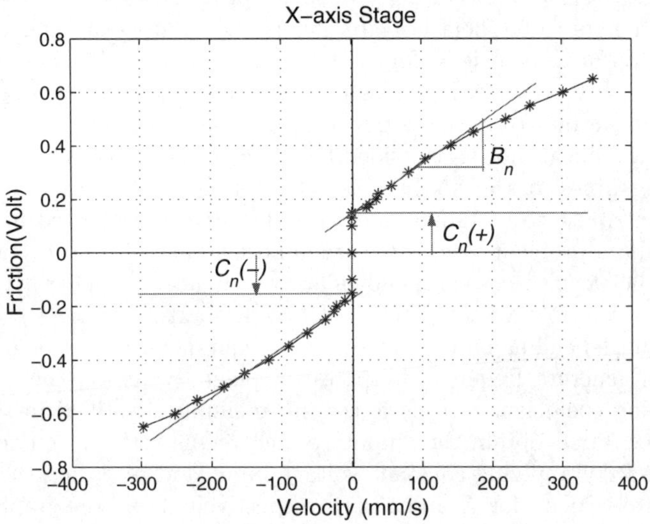


**Fig. A.1.** Friction identification of XY table (X-axis).

## A.2 Experimental Setup for Micro-Teleoperation

In the microteleoperation described in Chapter 6, the coarse/fine dual-stage in Fig. 6.10 was used to determine the position of microgripper in the slave

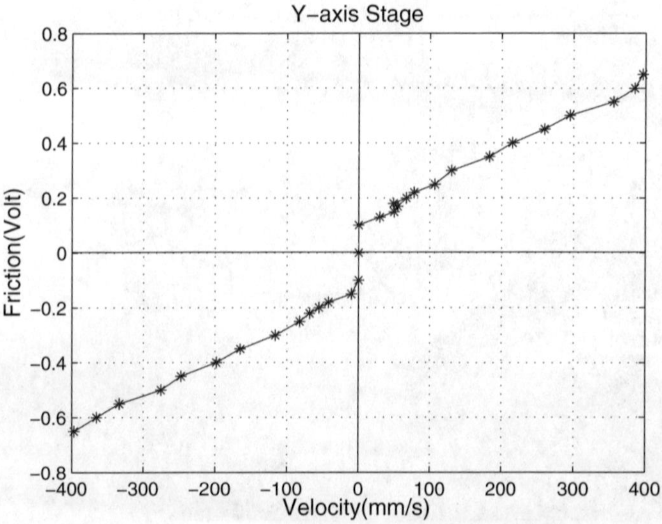


**Fig. A.2.** Friction identification of XY table (Y-axis).

system by following the reference trajectory from the master arm. The macro motion of the master device is scaled down usually as much as one over hundreds to one over thousands in the slave system so that the gripping tool attached at the end of the slave manipulator can perform microscopic tasks.

Regarding the microteleoperation system, it is composed of master arm (haptic device) [2, 89], micromanipulator [2, 89], microgripper [20, 88], microscope and vision system, etc. The purpose of microteleoperation system is to manipulate microscopic objects which are not distinguishable by human vision or to perform precission works which exceeds the resolution of human finger motion. The need of microteleoperation system is on an increasing trend in various applications, for example, bio cell manipulation for bio technology [108], tele-surgery in medical field, micro assembly for micromechanical systems [100], and others. However, still the micromanipulation system technology is not so matured. This is mainly because the physics of microscopic world has not been well understood and we don't have the technology to fabricate a reliable gripping tool to handle micro objects yet.

Figure A.3 is the experimental setup for microteleoperation which consists of 6-DOF master arm, 6-DOF slave manipulator (coarse/fine dual-stage and 3-DOF micromanipulator), micro gripper, and microscope and CCD camera. The diagram in Fig. A.4 denotes the bilateral communication between master and slave system. Figure A.5 shows the gripping tool (actuated by piezoelectric bender) is handling micro balls of about 80 $\mu$m diameter on the microscope table.

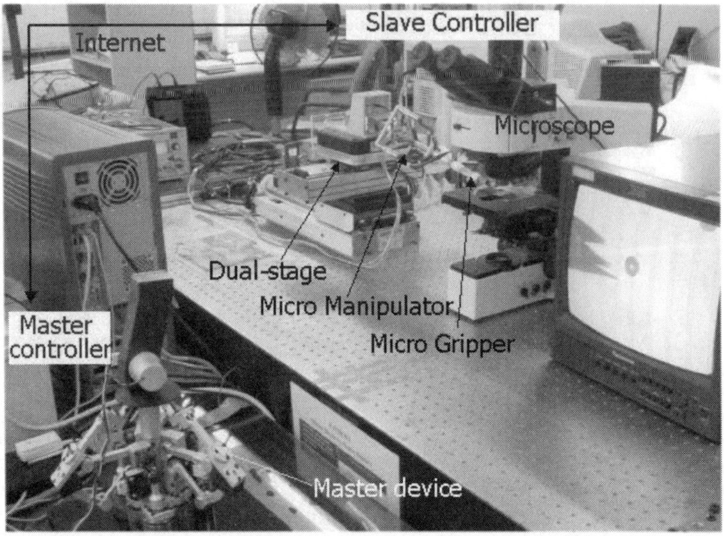


**Fig. A.3.** Experimental setup for microteleoperation.

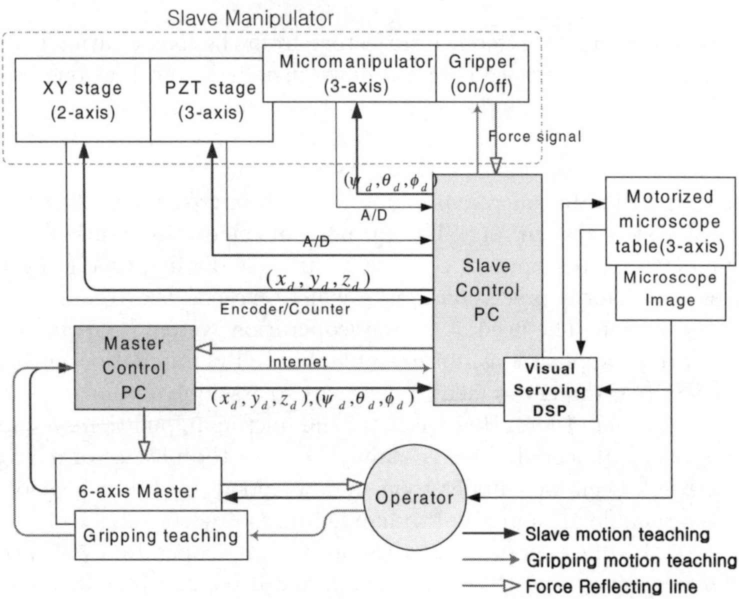


**Fig. A.4.** Bilateral communication lines in the microteleoperation system.

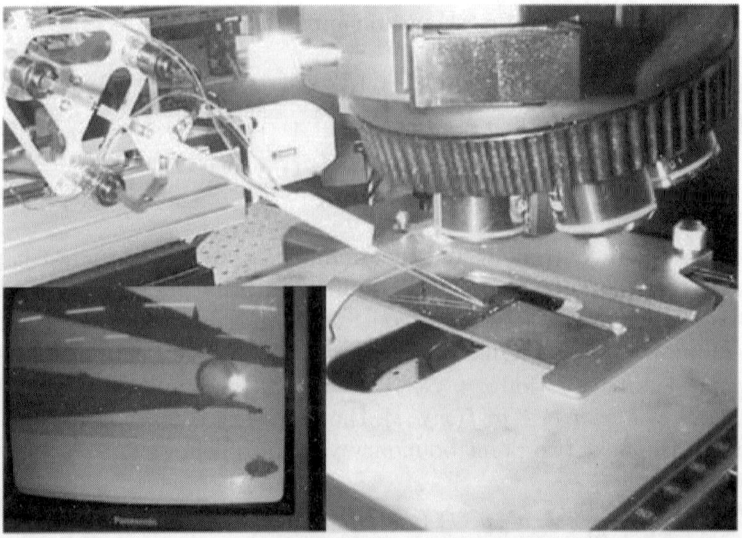

**Fig. A.5.** Handling of micro objects by gripping tool.

# A.3 Mathematical Theorems

## A.3.1 Matrix Inversion Lemma

In deriving the sensitivity functions in Chapter 5, the following inversion formula for a partitioned matrix was used.

Let a nonsingular and square matrix $A \in C^{(n+h)\times(n+h)}$ be partitioned as

$$A = \left[\begin{array}{c|c} A_{11} & A_{12} \\ \hline A_{21} & A_{22} \end{array}\right]$$

where $A_{11} \in C^{n\times n}$ and $A_{22} \in C^{h\times h}$ are square matrices. Then

$$\left[\begin{array}{c|c} A_{11} & A_{12} \\ \hline A_{21} & A_{22} \end{array}\right]^{-1} = \left[\begin{array}{c|c} A_{11}^{-1} + A_{11}^{-1}A_{12}\Delta^{-1}A_{21}A_{11}^{-1} & -A_{11}^{-1}A_{12}\Delta^{-1} \\ \hline -\Delta^{-1}A_{21}A_{11}^{-1} & \Delta^{-1} \end{array}\right]$$

with $\Delta \triangleq A_{22} - A_{21}A_{11}^{-1}A_{12}$ and also

$$\left[\begin{array}{c|c} A_{11} & A_{12} \\ \hline A_{21} & A_{22} \end{array}\right]^{-1} = \left[\begin{array}{c|c} \hat{\Delta}^{-1} & -\hat{\Delta}^{-1}A_{12}A_{22}^{-1} \\ \hline -A_{22}^{-1}A_{21}\hat{\Delta}^{-1} & A_{22}^{-1} + A_{22}^{-1}A_{21}\hat{\Delta}^{-1}A_{12}A_{22}^{-1} \end{array}\right]$$

with $\hat{\Delta} \triangleq A_{11} - A_{12}A_{22}^{-1}A_{21}$.

These relationships can be derived by letting $A^{-1} = \left[\begin{array}{c|c} W & X \\ \hline Y & Z \end{array}\right]$ and using

$$\left[\begin{array}{c|c} A_{11} & A_{12} \\ \hline A_{21} & A_{22} \end{array}\right]\left[\begin{array}{c|c} W & X \\ \hline Y & Z \end{array}\right] = \left[\begin{array}{c|c} I_n & 0 \\ \hline 0 & I_h \end{array}\right]$$

where the following identities should be applied.

$$(A_{11} - A_{12}A_{22}^{-1}A_{21})^{-1} = A_{11}^{-1} + A_{11}^{-1}A_{12}(A_{22} - A_{21}A_{11}^{-1}A_{12})^{-1}A_{21}A_{11}^{-1}$$
$$\text{and } (A_{22} - A_{21}A_{11}^{-1}A_{12})^{-1} = A_{22}^{-1} + A_{22}^{-1}A_{21}(A_{11} - A_{12}A_{22}^{-1}A_{21})^{-1}A_{12}A_{22}^{-1}.$$

### A.3.2 Pontrygin's Minimum Principle

The time-optimal control law and the minimum-time trajectory in Chapter 6 can be derived based on the following minimum principle [7, 43].

As is well-known, the optimal control problem is to find the control $u^*$ which minimizes the cost function, $J = h(t_f, x_f) + \int_{t_0}^{t_f} L(x, u, t)dt$ with the constraint of state equation $\dot{x} = f(x, u, t)$. When the Hamiltonian is defined as $H(x, u, p, t) = L(x, u, t) + p^T f(x, u, t)$, the necessary condition for optimality is given by the next two point boundary value problem:

$$\dot{x}^* = \frac{\partial H}{\partial p}(x^*, u^*, p^*, t) \text{ (state equation)}$$
$$\dot{p}^* = -\frac{\partial H}{\partial x}(x^*, u^*, p^*, t) \text{ (costate equation)}$$
$$x(t_0) = x_0, \quad \left[\frac{\partial h}{\partial x}(t_f) - p^*(t_f)\right]^T \delta x_f + \left[H^* + \frac{\partial h}{\partial t}(t_f)\right] \delta t_f = 0. \text{ (B.C.)s}$$

When the admissible controls and states are constrained by any boundaries, the following inequality condition should be added to the above necessary condition.

$$H^*(x^*, u^*, p^*, t) \leq H(x^*, u, p^*, t),$$

which indicates that *an optimal control minimizes the Hamiltonian (Pontrygin's Minimum Principle)*.

In time-optimal control problem, the Hamiltonian is given by $H(x, u, p, t) = 1 + p^T f(x, u, t)$. When we assume the plant state equation of the form $\dot{x} = a(x, u, t) + Bu$, the inequality condition of the Hamiltonian yields the optimal control $u^*(t) = -\text{sgn}(B^T p^*(t))$ and it is claimed that there exists a switching function $\mathcal{S}(t)$ such that $u^*(t) = -\text{sgn}(\mathcal{S}(t))$. Actually, finding of the switching function is the heart of the time-optimal control problem.

# References


1. Adriaens, Han J. M. T. A., de Koning, W. L., and Banning, R., 2000, "Modeling of Piezoelectric Actuators," *IEEE/ASME Trans. on Mechatronics*, Vol. 5, No. 4, Dec. pp. 331–341.
2. Ando, N., Korondi, P., and Hashimoto, H., 2001, "Development of Micomanipulator and Haptic Interface for Networked Micromanipulation," *IEEE/ASME Trans. on Mechatronics*, Vol. 6, No. 4, Dec. pp. 417–427.
3. Armstrong-Helouvry, B., 1991, *Control of Machines with Friction*, Kluwer Academic Publishers.
4. Armstrong-Helouvry, B., 1994, "A Survey of Models, Analysis Tools and Compensation Methods for the Control of Machines with Friction," *Autimatica*, Vol. 30, No. 7, pp 1083-1138.
5. Asada, H. and Slotine. J. -J. E., 1986, "Robot Anaysis and Control," *John Wiley and Sons.*
6. Astrom, K. J. and Wittenmark. B., 1997, *Computer-Controlled Systems*, 3rd ed., Prentice-Hall.
7. Athans, M and Falb, F. L., 1966, "Optimal control," Chapter 6-7, *McGraw-Hill.*
8. Baril, C., G. and Gutman, P.-O., 1997, "Performance Enhancing Adaptive Friction Compensation for Uncertain Systems," *IEEE Trans. Contr. Sys. Tech.*, Vol. 5, No. 5, Sep., pp. 466-479.
9. Bickel, R. and Tomizuka, M., 1999, "Passivity-Based Versus Dusturbance Observer Based Robot Control: Equivalence and Stability," *ASME J. of Dyn. Sys. Meas. and Cont.*, Vol. 121, Mar., pp. 41-47.
10. Chang, P. H., Lee, J. W., and Park, S. H., 1997, "Time Delay Observer: A Robust Observer for Nonlinear Plants," *ASME J. Dyn. Sys., Meas., and Contr.*, Vol. 119, Sep., pp. 521–527.
11. Chen, B. M., Lee, T. H., Hang, C.-C., Guo, Y., and Weerasooriya, S., 1999, "An $H_\infty$ Almost Disturbance Decoupling Robust Controller Design for a Piezoelctric Bimorph Actuator with Hysteresis," *IEEE Trans. on Cont. Sys. Tech.*, Vol. 7, No. 2, Mar. pp. 160-174.
12. Chiu, G. T.-C. and Tomizuka, M., 1998, "Coordinated Position Control of Multi-Axis Mechanical Systems," *ASME J. Dyn. Sys., Meas., and Contr.*, Vol. 120, Sep. pp. 389-393.

13. Choi, B.-K., Choi, C.-H., and Lim, H., 1999, "Model-Based Disturbance Aattenuation for CNC Machining Centers in Cutting Process," *IEEE/ASME Trans. on Mechatronics*, Vol.4, No.2, pp.157-168.
14. Choi, Y., Chung, W. K., and Suh, I. H., 2001, "Performance and $H_\infty$ Optimality of PID Trajectory Tracking Controller for Lagrangian Systems," *IEEE Trans. on Robot. and Auto.*, Vol. 17, No. 6, Dec., pp. 857-869
15. de Wit, C., Olsson, H., Astrom, K. J., and Lischinsky, P., 1995, "A New Model for Control of Systems with Friction," *IEEE Trans. on Auto. Contr.*, Vol. 40, No. 3, Mar., pp. 419-425.
16. Doyle, J. C. and Stein, G., 1981, "Multivariable Feedback Design: Concepts for a classical/Modern Synthesis," *IEEE Trans. Automat. Contr.*, Vol. AC-26, Feb., pp. 4-16.
17. Doyle, J. C. , Francis, B. A., and Tannenbaum, A. R., 1992, *Feedback Control Theory*, Maxwell Macmillan.
18. Elmali, H., and Olgac, N., 1992, "Sliding Mode Control with Perturbation Estimation," *Int. J. Control*, Vol. 56, pp. 923-941.
19. Eun, Y., Kim, J.-H., Kim, K., and D.-I. Cho, 1999, "Discrete-Time Variable Structure Controller with a Decoupled Disturbance Compensator and Its Application to a CNC Servomechanism", *IEEE Trans. on Cont. Sys. Tech.*, Vol. 7, No. 4, Jul., pp. 414-423.
20. Fahlbusch, St. and Fatikow, S., 2001, "Micro Force Sensing in a Micro Robotic System," *Proc. of 2001 IEEE Int. Conf. on Robot. and Auto.*, pp. 3435–3440.
21. Fan, L.-S. et. al., 1995, "Magnetic Recording Head Positioning at Very High Track Densities Using a Microactuator-Based, Two-Stage Servo System," *IEEE Trans. on Indust. Elec.*, Vol. 42, No. 3, Jun.,pp. 222-233.
22. Franklin, G. F., Powell, J. D., and Emami-Naeini, A., 1986, *Feedback control of dynamic systems*, Addison-Wesley.
23. Franklin, G. F., Powell, J. D., and Workman, M. L., 1990, *Digital control of dynamic systems*, 2nd ed., Addison-Wesley.
24. Friedland, B. and Park, Y.-J., 1992, "On Adaptive Friction Compensation," *IEEE Trans. on Auto. Contr.*, Vol. 37, No. 10, Oct., pp. 1609-1612.
25. Friedland, B., 1996, *Advanced Control System Design*, Prectice Hall.
26. Grewal, M. S. and Andrews, A. P., 1993, *Kalman Filtering: Theory and Practice*, Prentice-Hall.
27. Gu, D.-W. and Poon, F. W., 2001, "A Robust State Observer Scheme," *IEEE Trans. on Automatic Control*, Vol. 46, No. 12, Dec., pp. 1958–1963.
28. Haessig, D. A. and Friedland, B., 1991, "On the Modeling and Simulation of Friction," *ASME J. of Dyn. Sys. Meas. and Cont.*, Vol. 113, Sep., pp. 354-362.
29. Hernandez, D., Park, S.-S., Horowitz, R., and Packard, A. K., 1999, "Dual-Stage Track-Following Servo Design for Hard Disk Drives," *Proc. of 1999 American Contr. Conf.*, pp. 4116–4121.
30. Hollis, R. L., Salcudean, S. E. and Allan, A. P., 1991, "A Six-Degree-of-Freedom Magnetically Levitated Variable Compliance Fine-Motion Wrist: Design, Modeling, and Control," *IEEE Trans. on Robot. and Auto.*, Vol. 7, No. 3, pp. 320–332.
31. Hori, Y., 1993, "Robust Motion Control Based on A Two-Degree-of-Freedom Servosystem", *Advanced Robotics*, Vol. 7, No. 6, pp. 525-546.
32. Hou, M., Pugh, A. C., and Muller, P. C., 1999, "Disturbance Decoupled Functional Observers," *IEEE Trans. on Automatic Control*, Vol. 44, No. 2, Feb., pp. 382–386.

33. Hsia, T. C., 1989, "A New Technique for Robust Control of Servo Systems," *IEEE Trans. on Indust. Elec.*, Vol. 36, No. 1, Feb., pp. 1-7.
34. Hsia, T. C. and Gao, L. S., 1990, "Robot Manipulator Control Using Decentralized Linear Time-Invariant Time-Delayed Controllers," *1990 IEEE Int. Conf. on Robot. and Auto.*, pp. 2070-2075.
35. Huang S.-J., Yen, J.-Y., and Lu, S.-S., 1999, "Dual Mode Control of of a System with Friction," *IEEE Trans. on Cont. Sys. Tech.*, Vol. ,No. 3, May., pp. 306-314.
36. Iwasaki, M., Shibata, T., and Matsui, N., 1999, "Disturbance-Observer-Based Nonlinear Friction Compensation in Table Drive System," *IEEE/ASME Trans. on Mechatronics*, Vol. 4, No. 1, Mar., pp. 3-8.
37. Jeong, H.-S. and Lee, C.-W., 1997, "Time Delay Control With State Feedback for Azimuth Motion of the Frictionless Positioning Device," *IEEE/ASME Trans. on Mechatronics*, Vol. 2, No. 3, Sep. pp. 161–168.
38. Kalman, R. E., 1960, "A New Approach ti Linear Filtering and Prediction Problems," *Trans. of the ASME, Journal of Basic Engineering*, Mar., pp. 35–45.
39. Kempf, C. J. and S. Kobayashi, S., 1999, "Disturbance Observer and Feedforeard design for a High-Speed Direct-Drive Positioning Table," *IEEE Trans. on Contr. Sys. Tech.*, Vol. 7, No. 5, Sep., pp. 513–526.
40. Khatib, O. , 1995, "Inertial properties in robotic manipulation: An object-level framework," *Int. J. of Robot. Res.*, Vol. 13, No. 1, Feb. 1995, pp. 19-36.
41. Kim, B. K., Choi, H. T., Chung, W. K., and Suh, I. H., 2001, "Analysis and Design of Robust Motion Controllers in the Unified Framework," *ASME J. Dyn. Sys., Meas., and Contr.*, Vol. 124, Jun., pp. 313–321.
42. Kim, J. H. and Oh, J.-H., 2000, "Robust state estimator of stochastic linear systems with unknown disturbances," *IEE Proc. Control Theory Appl.*, Vol. 147, No. 2, Mar., pp. 224–228.
43. Kirk, D. E., 1970, "Optimal control theory," *Prentice Hall.*
44. Komada, S., Ohnishi, K., and Hori, T., 1991, "Hybrid Position/Force Control of Robot Manipulators Based on Accelearation Controller", *Proc. of the 1991 IEEE Int. Conf. on Robot. and Auto.* , pp. 48–55, 1991.
45. Komada, S., Machii, N., and Hori, T., 2000, "Control of Redundant Manipulators Considering Order of Disturbance Observer", *IEEE Trans. on Indust. Elec.*, Vol. 47, No. 2, Apr. pp. 413–420.
46. Koren, Y., 1980, "Cross-Coupled Biaxial Computer Control for Manufacturing Systems," *ASME J. Dyn. Sys., Meas., and Contr.*, Vol. 102, Dec. 1980, pp. 265–272.
47. Kulkarni, P. K. and Srinivasan, K., 1989, "Optimal Contouring Control of Multi-Axial Feed Drive Servomechanisms," *ASME J. Engineering for Industry*, Vol. 111, May. 1989, pp. 140–148.
48. Kwon, S. J., Chung, W. K., and Youm, Y., 2001, "A Novel Robust Tracking Controller Based on Hierarchical Perturbation Compensation," *Proc. of 2001 American Contr. Conf. (ACC)*, pp. 541-546.
49. Kwon, S. J., Chung, W. K., and Youm, Y., 2000, "Robust and Time-Optimal Control Strategy for Coarse/Fine Dual Stage Manipulators," *Proc. of 2000 IEEE Int. Conf. on Robot. and Auto. (ICRA)*, pp. 4051-4056.
50. Kwon, S. J., Chung, W. K., and Youm, Y., 2001, "On the Coarse/Fine Dual Stage Manipulators with Robust Perturbation Compensator," *Proc. of 2001 IEEE Int. Conf. on Robot. and Auto. (ICRA)*, pp. 121-126.

51. Kwon, S. J. and Chung, W. K., 2002, "A Discrete-Time Design and Analysis of Perturbation Observer," *Proc. of 2002 American Contr. Conf. (ACC)*, Alaska, USA.
52. Lawrence, D. A., 1993, "Stability and Transparency in Bilateral Teleoperation," *IEEE Trans. on Robot. and Auto.*, Vol. 9, No. 5, Oct., pp. 624-637.
53. Lee, C.W. and Kim, S.-W., 1997, "An Utraprecision stage for Alignment of wafers in Advanced microlithography," *Precision Engineering*, Vol. 21, Sep., pp. 113-122.
54. Lee, H. S., 1994, "Robust Digital Tracking Controllers for High-Speed/High-Accuracy Positioning Systems", Ph.D. Dissertation, UC Berkeley.
55. Lee, H. S. and Tomizuka, M., 1996, "Robust Motion Controller Design for High-Accuracy Positioning Systems", *IEEE Trans. on Indust. Elec.*, Vol. 43, Feb., pp. 48-55.
56. Lewis, F. L., 1992, *Applied Optimal Control and Estimation*, Prentice-Hall.
57. Li, Y. and Horowitz, R., 2000, "Track Following controller Design of MEMS Based Dual-Stage Servos in Magnetic Hard Disk Drives," Proc. of *2000 IEEE Int. Conf. on Robot. and Auto.*, pp. 953-958.
58. Luttrell, D. E., and Dow, T. A., 1988, "Control of Precise Positioning System with Cascaded Colinear Actuators," *Proc. of 1988 American Contr. Conf.*, pp.121-126.
59. Maciejowski, J. M., 1989, *Multivariable Feedback Design*, Addison-Wesley.
60. Masmoudi, R. A. and Hedrick, J. K., 1992, "Estimation of Vehicle Shaft Torque Using Nonlinear Observers," *ASME J. Dyn. Sys., Meas., and Contr.*, Vol. 114, Sep., pp. 394–400.
61. McCormick, J. and Horowitz, R., 1991, "Time optimal seek trajectories for a dual stage optical disk drive actuator," *ASME J. Dyn. Sys., Meas., and Contr.*, Vol. 113, Sep. 1991, pp. 534-536.
62. Misawa, E. A. and Hedrick, J. K., 1989, "Nonlinear Observers – A State of Art Survey," *ASME J. Dyn. Sys., Meas., and Contr.*, Vol. 111, Sep.
63. Mita, T., Hirata, M., Murata, K., and Zhang, H., 1998, "$H_\infty$ Control Versus Disturbance-Observer-Based Control," *IEEE Trans. on Indust. Elec.*, Vol. 45, No. 3, Dec. pp. 488–495.
64. Mittal, S. and Menq, C.-H., 1997, "Precision Motion Control of a Magnetic Suspension Actuator Using a Robust Nonlinear Compensation Scheme," *IEEE/ASME Trans. on Mechatronics*, Vol. 2, No. 4, Dec. pp. 268–280.
65. Morari, M. and Zafiriou, E., 1989, *Robust Process Control*, Prentice-Hall.
66. Morgan,R. G. and Ozguner, U., 1985, "A Decentralized Variable Structure Control Algorithm for Robotic Manipulators," *IEEE J. Robot. and Auto.*, Vol. RA-1, No. 1, Mar., pp. 57–65.
67. Morris, K., 2001, *Introducton to Feedback Control*, Academic Press.
68. Moura, J. T., Roy, R. G., and Olgac, N., 1997, "Frequency-Shaped Sliding Modes: Analysis and Experiments", *IEEE Trans. on Cont. Sys. Tech.*, Vol. 5, No. 4, Jul., pp. 394-401.
69. Moura, J. T., Elmali, H., and Olgac, N., 1997, "Sliding Mode Control with Sliding Perturbation Observer," *ASME J. Dyn. Sys., Meas., and Contr.*, Vol. 119, Dec., pp. 657–665.
70. Oh, Y. and Chung, W. K., 1999, "Disturbance Observer based Motion Control of Redundant Manipulators Using Inertially Decoupled Dynamics," *IEEE/ASME Trans. on Mechatronics*, Vol.4, No.2, pp.133-146.

71. Ohnishi, K., Shibata, M., and Murakami, T., 1996, "Motion Control for Advanced Mechatronics," *IEEE/ASME Trans. on Mechatronics*, Vol. 1, No. 1, Mar. pp. 56-67.
72. Ohishi, K., Ohnishi, K., and Miyachi, K., "Adaptive DC Servo Drive Control Taking Force Disturbance Suppression into Account," *IEEE Trans. on Industry Applications*, Vol. 24, No. 1, pp. 171-176, 1988.
73. Park, J., Chung, W. K., 2000, "Design of a Robust $H_\infty$ PID Control for Robot Manipulators," *ASME J. Dyn. Sys., Meas., and Contr.*, Vol. 122, Dec..
74. Park, J. H. and Kim, Y. M., 1999, "Time-Delay Sliding Mode Control for a Servo," *ASME J. Dyn. Sys., Meas., and Contr.*, Vol. 121, Mar., pp. 143-148.
75. Petersen, I. R. and McFarlane D. C, 1994, "Optimal Guaranteed Cost Control and Filtering for Uncertain Linear Systems," *IEEE Trans. on Automatic Control*, Vol. 39, No. 9, Sep., pp. 1971–1977.
76. Ro, P. I. and Hubbel, P. I., 1993, "Model Reference Adaptive Control of Dual-Mode Micro/Macro Dynamics of Ball Screws for Nanometer Motion," *ASME J. of Dyn. Sys. Meas. and Cont.*, Vol. 115, Mar., pp. 103-108.
77. Sadegh, N. and Horowitz, R., 1990, "Stability and Robustness Analysis of a Class of Adaptive Controllers for Robot Manipulators," *Int. J. of Robotics Research*, Vol. 9, No. 3, pp 74-92.
78. Salcudean, S. and An, C., 1989, "On the control of redundant coarse-fine manipulators," *Proc. of 1989 IEEE Int. Conf. on Robot. and Auto.*, pp. 1834–1840.
79. Schroeck, S. J. and Messner, W. C., 1999, "On Controller Design For Linear-Time-Invariant Dual-Input Single-Output Systems," *Proc. of 1999 American Contr. Conf.*, pp. 4121–4126.
80. Shaked, U. de Souza, C. E., 1995, "Robust Minimum Variance Filtering," *IEEE Trans. on Signal Processing*, Vol. 43, No. 11, Nov., pp. 2474–2483.
81. Shahian, B. and Hassul, M., 1993, "Control System Design Using MATLAB," *Prentice Hall.*
82. Sharon, A., Hogan, N., and Hardt, D. E., 1988, "High Bandwidth Force Regulation and Inertia Reduction Using a Macro/Micro Manipulator System," *Proc. of 1988 IEEE Int. Conf. on Robot. and Auto.*, pp. 126–132.
83. Sharon, A., Hogan, N., and Hardt, D. E., 1993, "The macro/micro manipulator: An improved architecture for robot control," *Robotics and Computer Integrated Manufacturing*, Vol. 10, No. 3, pp. 209-222.
84. Slotine, J.-J. E., Hedricj, J. K., and Misawa, E. A., 1987, "Sliding Mode Control with Sliding Perturbation Observer," *ASME J. Dyn. Sys., Meas., and Contr.*, Vol. 109, Sep., pp. 245–252.
85. Slotine, J.-J. E. and Li, W., 1991, "Applied Nonlinear Control," *Prentice Hall.*
86. Slotine, J-J. E. and Li, W., 1987, "On the adaptive Control of Robot Manipulators," *Int. J. of Robotics Research*, Vol. 6, No. 3, pp 49-59.
87. Staroselsky, S. and Stelson, K. A., 1988, "Two-Stage Actuation for Improved Accuracy of Contouring," *Proc. of 1988 American Contr. Conf.*, pp. 127-132.
88. Tanikawa, T. et al., 2001, "Force Control System for Autonomous Micro Manipulation," *Proc. of 2001 IEEE Int. Conf. on Robot. and Auto.*, pp. 610–615.
89. Tanikawa, T. and Arai, T., 1999, "Development of a Mico-Manipulation System Having a Two-Fingered Micro-Hand" *IEEE Trans. on Robot. and Auto.*, Vol. 15, No. 1, Feb. pp. 152–162.
90. Tesfaye, A. and Tomizuka, M., 1995, "Robust Control of Discretized Continuous Systems Using the Theory of Sliding Modes," *Int. J. Contr.* , Vol. 62, pp. 209–226.

91. Tesfaye, A., Lee, H. S., and Tomizuka, M., 2000, "A sensitivity Optimization Approach to Design of a Disturbance Observer in Digital Motion Control Systems ," *IEEE/ASME Trans. on Mechatronics*, Vol. 5, No. 1, pp.32-38.
92. Tomizuka, M., 1987, " Zero Phase Error Tracking Algorithm for Digital Control," *ASME J. Dyn. Sys., Meas., and Contr.*, Vol. 109, Mar., pp. 65-68.
93. Tomizuka, M., Tsao, T.-C., and Chew, K.-K., 1989, "Analysis and Synthesis of Discrete-time Repetitive Controllers", *ASME J. Dyn. Sys., Meas., and Contr.*, Vol 111, Sep., pp 353-358.
94. Tomizuka, M., 1993, "On the Design of Digital Tracking Controllers," *ASME J. Dyn. Sys., Meas., and Contr.*, Vol. 115(B), Jun., pp. 412-418.
95. Torfs, D., De Schutter, J., and Swevers, J., 1992, "Extended Bandwidth Zero Phase Error Tracking Control of Nonminimal Phase System," *ASME J. Dyn. Sys., Meas., and Contr.*, Vol. 114, Sep., pp. 347-351.
96. Tsui. C.-C., 1996, "A New Design Approach to Unknown Input Observers," *IEEE Trans. on Automatic Control*, Vol. 41, No. 3, Mar., pp. 464-468.
97. Umeno, T. and Hori, Y., 1991, "Robust Speed Control of DC Servomotors Using Modern Two Degree-of-Freedom Controller Design", *IEEE Trans. on Indust. Elec.*, Vol. 38, No. 5, Oct. pp. 363-368.
98. Walcott, B. L. and Zak, S. H., 1988, "Combined Observer–Controller Synthesis for Uncertain Dynamical Systems with Applications," *IEEE Trans. on Systems, Man, and Cybernetics*, Vol. 18, Feb., pp. 88–104.
99. Widraw, B. and Walach, E., 1996, *Adaptive Inverse Control*, Prectice Hall.
100. Yang, G. et al., 2001, "A Flexible Experimental Workcell for Efficient and Reliable Wafer-Level 3D Microassembly," *Proc. of 2001 IEEE Int. Conf. on Robot. and Auto.*, pp. 133–138.
101. Yao, B., Al-Majed, M., and Tomizuka, M., 1997, "High-Performance Robust Motion Control of Machine Tools: An Adaptive Robust Control Approach and Comparative Experiments," *IEEE/ASME Trans. on Mechatronics*, Vol. 2, No. 2, Jun. pp. 63-76.
102. Yeh, S.-S. and Hsu, P.-L., 1999, "Analysis and Design of the Integrated Controller for Precise Motion Systems," *IEEE Trans. on Cont. Sys. Tech.*, Vol. 7, No. 6, Nov. pp. 706–717.
103. Yen, J.-Y., Hallamasek, K., and Horowitz, R., 1990, "Track Following controller Design for a Compound Disk Drive Actuator," *ASME J. of Dyn. Sys. Meas. and Contr.*, Vol. 112, Sep., pp. 391-402.
104. Yoshikawa, S., Harada, K., and Matsumoto, A., 1996, "Hybrid position/force control of flexible-macro/rigid-micro manipulator systems," *IEEE Trans. on Robot. and Auto.*, Vol. 12, NO. 4, Aug. 1996, pp. 633-640.
105. Youcef-Toumi, K. and Ito, O., 1990, "A Time Delay Controller for Systems With Unknown Dynamics," *ASME J. Dyn. Sys., Meas., and Contr.*, Vol. 112, Mar., pp. 133-142.
106. Youcef-Toumi, K. and Reddy, S., 1992, "Analysis of Linear Time Invariant Systems With Time Delay," *ASME J. Dyn. Sys., Meas., and Contr.*, Vol. 114, Dec., pp. 544-555.
107. Young, K. D. and Ozguner, U., 1993, "Frequency shaping compensator design for sliding mode", *Int. J. of Control*, Vol. 57, No. 5, pp. 1005-1019.
108. Yu, S. and Nelson, B. J., 2001, "Microrobotic Cell Injection," *Proc. of 2001 IEEE Int. Conf. on Robot. and Auto.*, pp. 620–625.

# Lecture Notes in Control and Information Sciences

## Edited by M. Thoma and M. Morari

Further volumes of this series can be found on our homepage: springeronline.com

**2001–2004 Published Titles:**

**Vol. 257:** Moallem, M.; Patel, R.V.; Khorasani, K.
Flexible-link Robot Manipulators
176 p. 2001 [1-85233-333-2]

**Vol. 258:** Isidori, A.; Lamnabhi-Lagarrigue, F.; Respondek, W. (Eds)
Nonlinear Control in the Year 2000 Volume 1
616 p. 2001 [1-85233-363-4]

**Vol. 259:** Isidori, A.; Lamnabhi-Lagarrigue, F.; Respondek, W. (Eds)
Nonlinear Control in the Year 2000 Volume 2
640 p. 2001 [1-85233-364-2]

**Vol. 260:** Kugi, A.
Non-linear Control Based on Physical Models
192 p. 2001 [1-85233-329-4]

**Vol. 261:** Talebi, H.A.; Patel, R.V.; Khorasani, K.
Control of Flexible-link Manipulators Using Neural Networks
168 p. 2001 [1-85233-409-6]

**Vol. 262:** Dixon, W.; Dawson, D.M.; Zergeroglu, E.; Behal, A.
Nonlinear Control of Wheeled Mobile Robots
216 p. 2001 [1-85233-414-2]

**Vol. 263:** Galkowski, K.
State-space Realization of Linear 2-D Systems with Extensions to the General nD (n>2) Case
248 p. 2001 [1-85233-410-X]

**Vol. 264:** Baños, A.; Lamnabhi-Lagarrigue, F.; Montoya, F.J
Advances in the Control of Nonlinear Systems
344 p. 2001 [1-85233-378-2]

**Vol. 265:** Ichikawa, A.; Katayama, H.
Linear Time Varying Systems and Sampled-data Systems
376 p. 2001 [1-85233-439-8]

**Vol. 266:** Stramigioli, S.
Modeling and IPC Control of Interactive Mechanical Systems – A Coordinate-free Approach
296 p. 2001 [1-85233-395-2]

**Vol. 267:** Bacciotti, A.; Rosier, L.
Liapunov Functions and Stability in Control Theory
224 p. 2001 [1-85233-419-3]

**Vol. 268:** Moheimani, S.O.R. (Ed)
Perspectives in Robust Control
390 p. 2001 [1-85233-452-5]

**Vol. 269:** Niculescu, S.-I.
Delay Effects on Stability
400 p. 2001 [1-85233-291-316]

**Vol. 270:** Nicosia, S. et al.
RAMSETE
294 p. 2001 [3-540-42090-8]

**Vol. 271:** Rus, D.; Singh, S.
Experimental Robotics VII
585 p. 2001 [3-540-42104-1]

**Vol. 272:** Yang, T.
Impulsive Control Theory
363 p. 2001 [3-540-42296-X]

**Vol. 273:** Colonius, F.; Grüne, L. (Eds)
Dynamics, Bifurcations, and Control
312 p. 2002 [3-540-42560-9]

**Vol. 274:** Yu, X.; Xu, J.-X. (Eds)
Variable Structure Systems: Towards the $21^{st}$ Century
420 p. 2002 [3-540-42965-4]

**Vol. 275:** Ishii, H.; Francis, B.A.
Limited Data Rate in Control Systems with Networks
171 p. 2002 [3-540-43237-X]

**Vol. 276:** Bubnicki, Z.
Uncertain Logics, Variables and Systems
142 p. 2002 [3-540-43235-3]

**Vol. 277:** Sasane, A.
Hankel Norm Approximation for Infinite-Dimensional Systems
150 p. 2002 [3-540-43327-9]

**Vol. 278:** Chunling D. and Lihua X. (Eds)
$H_\infty$ Control and Filtering of Two-dimensional Systems
161 p. 2002 [3-540-43329-5]

**Vol. 279:** Engell, S.; Frehse, G.; Schnieder, E. (Eds)
Modelling, Analysis, and Design of Hybrid Systems
516 p. 2002 [3-540-43812-2]

**Vol. 280:** Pasik-Duncan, B. (Ed)
Stochastic Theory and Control
564 p. 2002 [3-540-43777-0]

**Vol. 281:** Zinober A.; Owens D. (Eds)
Nonlinear and Adaptive Control
416 p. 2002 [3-540-43240-X]

**Vol. 282:** Schröder, J.
Modelling, State Observation and Diagnosis of Quantised Systems
368 p. 2003 [3-540-44075-5]

**Vol. 283:** Fielding, Ch. et al. (Eds)
Advanced Techniques for Clearance of Flight Control Laws
480 p. 2003 [3-540-44054-2]

**Vol. 284:** Johansson, M.
Piecewise Linear Control Systems
216 p. 2003 [3-540-44124-7]

**Vol. 285:** Wang, Q.-G.
Decoupling Control
373 p. 2003 [3-540-44128-X]

**Vol. 286:** Rantzer, A. and Byrnes C.I. (Eds)
Directions in Mathematical Systems Theory and Optimization
399 p. 2003 [3-540-00065-8]

**Vol. 287:** Mahmoud, M.M.; Jiang, J. and Zhang, Y.
Active Fault Tolerant Control Systems
239 p. 2003 [3-540-00318-5]

**Vol. 288:** Taware, A. and Tao, G.
Control of Sandwich Nonlinear Systems
393 p. 2003 [3-540-44115-8]

**Vol. 289:** Giarré, L. and Bamieh, B.
Multidisciplinary Research in Control
237 p. 2003 [3-540-00917-5]

**Vol. 290:** Borrelli, F.
Constrained Optimal Control of Linear and Hybrid Systems
237 p. 2003 [3-540-00257-X]

**Vol. 291:** Xu, J.-X. and Tan, Y.
Linear and Nonlinear Iterative Learning Control
189 p. 2003 [3-540-40173-3]

**Vol. 292:** Chen, G. and Yu, X.
Chaos Control
380 p. 2003 [3-540-40405-8]

**Vol. 293:** Chen, G. and Hill, D.J.
Bifurcation Control
320 p. 2003 [3-540-40341-8]

**Vol. 294:** Benvenuti, L.; De Santis, A.; Farina, L. (Eds)
Positive Systems: Theory and Applications (POSTA 2003)
414 p. 2003 [3-540-40342-6]

**Vol. 295:** Kang, W.; Xiao, M.; Borges, C. (Eds)
New Trends in Nonlinear Dynamics and Control, and their Applications
365 p. 2003 [3-540-10474-0]

**Vol. 296:** Matsuo, T.; Hasegawa, Y.
Realization Theory of Discrete-Time Dynamical Systems
235 p. 2003 [3-540-40675-1]

**Vol. 297:** Damm, T.
Rational Matrix Equations in Stochastic Control
219 p. 2004 [3-540-20516-0]

**Vol. 298:** Choi, Y.; Chung, W.K.
PID Trajectory Tracking Control for Mechanical Systems
127 p. 2004 [3-540-20567-5]

**Vol. 299:** Tarn, T.-J.; Chen, S.-B.; Zhou, C. (Eds.)
Robotic Welding, Intelligence and Automation
214 p. 2004 [3-540-20804-6]

**Vol. 300:** Nakamura, M.; Goto, S.; Kyura, N.; Zhang, T.
Mechatronic Servo System Control
Problems in Industries and their Theoretical Solutions
212 p. 2004 [3-540-21096-2]

**Vol. 301:** de Queiroz, M.; Malisoff, M.; Wolenski, P. (Eds.)
Optimal Control, Stabilization and Nonsmooth Analysis
373 p. 2004 [3-540-21330-9]

**Vol. 302:** Filatov, N.M.; Unbehauen, H.
Adaptive Dual Control: Theory and Applications
237 p. 2004 [3-540-21373-2]

**Vol. 303:** Mahmoud, M.S.
Resilient Control of Uncertain Dynamical Systems
278 p. 2004 [3-540-21351-1]

**Vol. 304:** Margaris, N.I.
Theory of the Non-linear Analog Phase Locked Loop
303 p. 2004 [3-540-21339-2]

**Vol. 305:** Nebylov, A.
Ensuring Control Accuracy
256 p. 2004 [3-540-21876-9]

**Vol. 306:** Bien, Z.Z.; Stefanov, D. (Eds.)
Advances in Rehabilitation
472 p. 2004 [3-540-21986-2]